# High Temperature Environmentally Sustainable Energy Processes

Topical Conference at the 2010 AIChE Annual Meeting

## Salt Lake City, Utah, USA
## 7-12 November 2010

ISBN: 978-1-61782-143-1

**Printed from e-media with permission by:**

Curran Associates, Inc.
57 Morehouse Lane
Red Hook, NY 12571

**Some format issues inherent in the e-media version may also appear in this print version.**



Printed by Curran Associates, Inc. (2011)



AIChE
3 Park Avenue
New York, NY 10016-5991

Phone: (203) 702-7660
Fax: (203) 775-5177

www.aiche.org

**Additional copies of this publication are available from:**

Curran Associates, Inc.
57 Morehouse Lane
Red Hook, NY 12571 USA
Phone: 845-758-0400
Fax: 845-758-2634
Email: curran@proceedings.com
Web: www.proceedings.com

# TABLE OF CONTENTS

# (321a) The Suitability of Natural Commodities as Oxygen Carriers in Chemical Looping Combustion

Suman K. Hajra and **Thomas O'Brien**, Simulation, Analysis & Computational, NETL/DOE, Morgantown, WV

Chemical-looping combustion (CLC) technology may play a vital role in mitigating carbon dioxide emissions from the power generation and industrial sectors. In CLC, oxygen is transferred from the combustion air to the fuel by an oxygen carrier, usually metal oxide particles, circulated between two reactors, an air and a fuel reactor. Hence, the combustion of a fuel is carried out without mixing the products of combustion, carbon dioxide and water, with the nitrogen from the combustion air. When utilizing solid fuels directly, both the fuel and the oxidizer are fed into the fuel reactor as solids. Heterogeneous reactions (devolatilization and gasification, e.g.) generate gaseous products which subsequently contact the metal oxides. The oxygen carrier particles play a prominent role in the success of CLC technique. The particles must be able to convert the fuel to carbon dioxide and water efficiently, both thermodynamically and kinetically. They must resistant fragmentation and attrition due to the fluidizing conditions and chemical and thermal cycling. For a commercial scale CLC process, the particles also need to be environmentally benign and inexpensive. We considered commodity materials, many related to iron ore, as possible oxygen carriers and review their sources, physical and chemical properties, cost, etc. A survey of these materials is presented to assist in the selection of oxygen carrier in the CLC technology. The study identified several promising commodity material for further evaluation as oxygen carrier particles for a CLC process.

See more of this Session: Chemical Looping Processes I
See more of this Group/Topical: Topical E: High Temperature Environmentally Sustainable Energy Processes (sessions joint with the Environmental Division)

# (321b) Investigation of Varying Isothermal and Isobaric Reaction Conditions for Copper Functioning as An Oxygen Carrier in Chemical Looping Combustion

**Richard Baraki**[1], Gabor Konya[2], Edward M. Eyring[2] and Adel. F. Sarofim[1], (1)Department of Chemical Engineering, University of Utah, Salt Lake City, UT, (2)Department of Chemistry, University of Utah, Salt Lake City, UT

The Cu/CuO system is a useful oxygen carrier in Chemical Looping Combustion (CLC). The oxidation and reduction reactions of these metal/metal oxide couples were investigated with thermo gravimetric analysis (TGA) under various isothermal and isobaric conditions. The results are used to determine the optimal operating conditions for a CLC system.

See more of this Session: Chemical Looping Processes I
See more of this Group/Topical: Topical E: High Temperature Environmentally Sustainable Energy Processes (sessions joint with the Environmental Division)

# INVESTIGATION ON REACTION MECHANISM OF CHEMICAL LOOPING COMBUSTION OF COAL UTILIZING OXYGEN CARRIERS

*Hanjing Tian*[1,2] *Ranjani Siriwardane*[1], *Duane Miller*[1,2], *George Richards*[1], *James Poston*[1] *and Thomas Simonyi*[1,2]

[1] U.S. DOE, National Energy Technology Labortory, 3610 Collins Ferry Road, P.O.Box 880, Morgantown, WV 26507-0880

[2] URS, 3610 Collins Ferry Road, Morgantown, WV 26505

## Introduction

Capturing carbon dioxide from power plants using fossil fuels is one of several strategies to reduce global carbon dioxide emissions. The task of removing $CO_2$ from power plant flue gas is challenging because existing methods to separate the $CO_2$ from the gas mixture requires a significant fraction of the power plant output. Various reviews describe emerging technologies to "scrub" $CO_2$ emitted from power plants or use oxy-fuel combustion. All these technologies require a significant energy input, so that alternate strategies to supply oxygen to the combustion are being investigated. Chemical looping combustion (CLC) is an emerging technology for clean energy production from fossil and renewable fuels. CLC produces sequestration-ready $CO_2$-streams without significant energy penalty[1-4]. A solid fuel such as coal is rarely used in CLC since the process with solid fuels faces many challenges.[5-8] The reaction mechanism of coal and metal oxides has not been well established. The generally accepted mechanism is that the coal de-volatilization products initiate the chain reaction with the metal oxide to complete the combustion reaction:[5-7]

$CO + MeO \rightarrow CO_2 + Me$ [1]

$CO_2 + C \rightarrow 2CO$ [2]

Other researchers have proposed utilization of oxygen from the metal oxide decomposition for combusting solid fuel (CLOU). This research work focused in understanding the reaction mechanism of solid fuels with oxygen carriers.

Thermo gravimetric studies, flow reactor studies, in-situ high temperature XRD, SEM and XPS studies were conducted with various solid fuels such as coal, coal char and carbon mixed with various metal oxides in the presence of both nitrogen and carbon dioxide. These studies indicated that the combustion reaction of carbon with metal oxides can occur at low temperatures (~500-800 $^0$C). Particles sizes and separation distance of the fuel and the oxygen carrier had a significant influence on the reaction temperature. Our experimental data suggest that presence of fuels in the vicinity of metal oxide may promote oxygen release from metal oxide when there is close contact between the fuel and metal oxide.

## Experimental

Solid fuels, including coal (Illinois #6), char, and carbon, were mixed physically with metal oxides in a Crescent WIG-BUG shaker. The mass ratio (metal oxide mass/coal mass) corresponds to stoichiometric oxygen supply. CuO and $Fe_2O_3$ (99%, Aldrich) were used as oxygen carriers with average particle sizes of 5 microns.

Thermogravimetric analysis (TGA) was conducted in a TA analyzer (Model 2050) with a mass spectrometer (Pfeiffer Thermostar D-35614) for analyzing the composition of the outlet gas stream from TGA. The samples were placed in a 5-mm deep, 10-mm diameter crucible. About 100 mg of the solid fuel-metal oxide mixture was heated in a quartz bowl from ambient to 900 or 1000 °C at a heating rate of 15 °C/min in various gases with flow rate of 100 sccm. The sample was then kept isothermal for about 30 minutes. The global rates of reactions (dX/dt) were calculated by differentiating fractional conversions (X).

In-situ XRD analyses were carried out using a Panalytical PW 3040 X-Pert Pro XRD system equipped with a 60 kV PW 3373/00 Cu LFF high-power ceramic tube with a Cu anode and a PW 3011/20 detector. The char/CuO mixture was heated in the hot stage up to 750 °C and XRD patterns were recorded for every 50 °C ramping.

X-ray photoelectron spectroscopy (XPS) was obtained with a Physical Electronics (PHI) model 32-096 X-ray source control and a 22-040 power supply interfaced to a model 04-548 X-ray source with an Omni Focus III spherical capacitance analyzer (SCA). SEM analysis is conducted with a JEOL 7600 FESEM system interfaced to a Thermo-Electron System 7 microanalysis system.

## Results and discussions

When coal was heated in the presence of a continuous flow of either $N_2$ or $CO_2$ in TGA without the metal oxide oxygen carrier present, a weight loss due to coal de-volatilization was observed around 439 °C. The coal de-volatilization reaction was complete around 600 °C. The combustion reaction of coal with various metal oxides all took place above 650 °C, at reaction rates substantially higher than that for the de-volatilization reaction. Among all oxygen carriers tested, CuO had the lowest combustion temperature, initiating around 650 °C with a rate maximum around 700°C, while coal CLC with $Fe_2O_3$ occurred at 566 and 980 °C as shown in Fig 1. When the combustion reaction with the metal oxides takes place above 650 °C, coal de-volatilization products were not present in TGA. Consequently, it is not possible for the coal de-volatilization products to initiate the combustion reaction.

For further investigation, pure carbon was used in similar experiments. When pure carbon was heated in a TGA test with a $N_2$ flow up to 900 °C, no weight changes were observed. When a mixture of carbon and CuO was heated in $N_2$, a combustion reaction with a significant rate was observed around 480 °C. Since volatiles are not produced from heating of carbon, the rapid reaction of CuO and carbon did not proceed via coal volatiles as described in the previous reported reaction mechanism.

Figure 1: Combustion performance of coal and $Fe_2O_3$

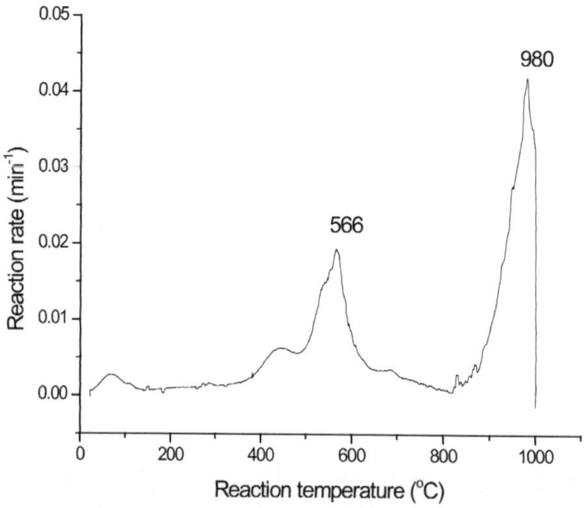

Other reports also have indicated that the reduction of metal oxides by carbon proceeds via the reaction with $CO_2$. In order to evaluate this, carbon was heated with pure $CO_2$. The major reaction with $CO_2$ and carbon was not observed up to 900 °C with no reaction around 500 °C. This indicates that the reaction of C/CuO that takes place at a fast rate around 480 °C does not necessarily occur via the reaction with $CO_2$. Thus, we conclude that the widely accepted mechanism as shown in reaction [2] cannot explain the present data.

Coal/metal oxide contact plays a key role on the reaction performance. CuO and carbon powders in four different configurations (Figure 2) were heated from ambient to 800 °C in a lab-scale fixed-bed reactor with a continuous flow of helium. A layer of quartz wool with average thickness of 0.25 inch was placed at the top and bottom of samples tubing, as well as to separate carbon and CuO. For sample (I)-C/CuO mixture, mass spectral (MS) data obtained from the outlet gas indicated the presence of significant amount of $CO_2$ and some CO, with the major peak around 495 °C. For sample (II)-CuO at the gas outlet location and sample (IV)-pure CuO, oxygen was observed around 800 °C due to CuO decomposition. For sample (III)-carbon at the gas outlet location, $CO_2$ and CO were detected around 800 °C due to the reaction of oxygen released from CuO with C. This confirms that when carbon and CuO are not in contact, the spontaneous decomposition of CuO occurs around 800 °C but only when CuO and C are in close contact, carbon can induce the oxygen release from CuO for combustion at a significantly lower temperature (480 °C). A series of flow reactor tests were also conducted with $Fe_2O_3$ and C pellets with same configuration up to 800 °C. Only the sample with C/$Fe_2O_3$ mixture produced $CO_2$ below than 800 °C as shown in Figure 3, and other configurations did not show any gaseous species below 800 °C since the decomposition temperature of $Fe_2O_3$ is higher than 800 °C .

Figure 2: Outlet gas composition and Temperature as a function of reaction time from flow reactor tests: sample (I)- carbon/CuO mixture, sample (II)- C and CuO separated by a 0.25" layer of quartz wool with CuO at the gas outlet location, sample (III)- C and CuO separated by a 0.25" layer of quartz wool with C at the gas outlet location, sample (IV) - CuO, MS signal *2,.

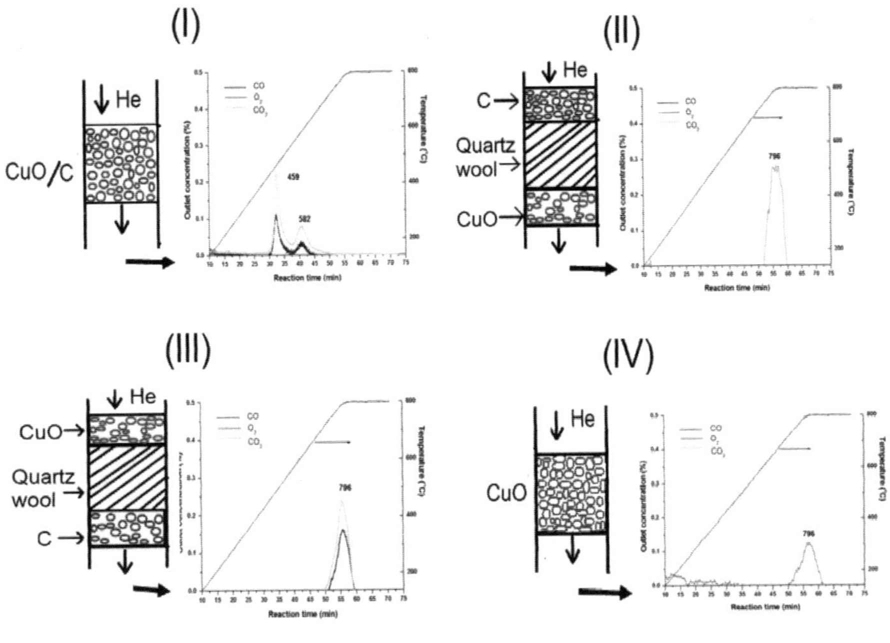

Figure 3: Combustion performance of carbon and $Fe_2O_3$ in a lab-scale fixed-bed reactor

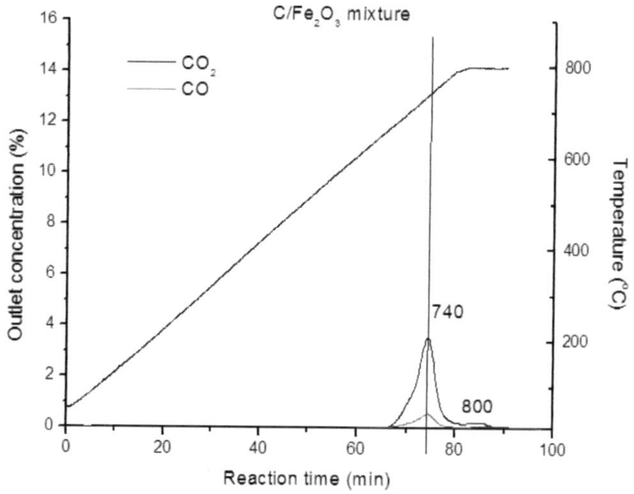


When CuO or $Fe_2O_3$ with various particle sizes were reacted with coal, the reaction temperatures increased with increasing particle size. Similarly, increase in coal particle size also resulted in increase in the reaction temperature. TGA reactions were also performed with carbon and CuO or $Fe_2O_3$ samples diluted with various amounts of quartz powder. The increased dilution by quartz powder increased the reaction temperatures. This further shows the importance of having carbon in the vicinity of the metal oxide to facilitate the oxygen release.

CLOU process proposed by some researchers utilizes oxygen released from metal oxide decomposition for combustion of solid fuel. However, the current observed combustion reaction of C/CuO and C/$Fe_2O_3$ mixtures take place around 560 and 800 °C, respectively. Thus, the CLOU process cannot be used for explaining the lower temperature reaction between metal oxide and C since $Fe_2O_3$ cannot decompose below 800 and1000 °C, which is confirmed by TGA.

In-situ X-ray diffraction and XPS data indicate that the carbon/metal oxide combustion through multiple reduction process: CuO is converted to $Cu_2O$ around 450 °C, and the reaction proceeds until $Cu_2O$ is converted to Cu metal, $Fe_2O_3$ is first reduced to FeO, and then fully reduced to Fe. . SEM images indicated that surface melting of the metallic copper layer formed initially on the CuO surface and continued reaction of C with the bulk CuO through the liquid layer.

**Reference**

1. H. Richter, K. Knoche, ACS Symp. Ser. 235, 71 (1983).
2. M. Ishida, H. Jin, J. Chem. Eng. Jpn. 27, 296, (1994).
3. M. Anheden, G. Svedberg, Energy Convers. Mgmt. 39, 1967(1998).
4. L. F. Diego et al., Fuel 83, 1749 (2004).
5. Y. Cao, B. Casenas, W. Pan, Energy and Fuels 20, 1845 (2006).
6. R. K. Lyon, J. A. Cole, Combustion and Flame 121, 249 (2000).
7. S. A. Scott, J. S. Dennis, A. N. Hay Hurst, AICHE Journal 52, 3325 (2006).
8. T. Mattison, A. Lyngfelt, H. Leion, Inter. J. Greenhouse Gas Control 3, 11 (2009).

# (321d) Syngas Production Via Chemical Looping

**Saurabh K. Bhavsar**[1], Michelle Najera[2], Rahul D. Solunke[1] and Götz Veser[1], (1)Chemical Engineering Department, University of Pittsburgh, US DOE-National Energy Technology Laboratory, Pittsburgh, Pittsburgh, PA, (2)Chemical Engineering Department, University of Pittsburgh, US DOE-National Energy Technology Laboratory, Pittsburgh; Mascaro Center for Sustainable Innovation, University of Pittsburgh, Pittsburgh, PA

Chemical looping combustion is an emerging clean combustion technology with inherent CO2 capture in which an oxygen carrier (typically a metal) is exposed to cyclic oxidation/reduction with air and a fuel, respectively. By changing the nature of the oxidant, the chemical looping principle can be extended onto related fuel processing reactions. Specifically, by using steam as oxidant, fuel-cell ready hydrogen streams can be obtained as the effluent of the oxidizer reactor. Similarly, replacing air with CO2 as oxidant, CO is obtained as product of the carrier oxidation process. Using methane as fuel for these processes, the first process results in net steam reforming of methane, while the second process results in dry reforming, although in both cases the selectivities are significantly different from the conventional processes. For both of these chemical looping reforming (CLR) processes, significant challenges lie in the slower oxidation kinetics when using steam or CO2 as oxidants, as well as the high-temperature and mechanical stability of the carriers. We are reporting on a combined experimental and theoretical investigation, in which a range of metals was evaluated as carriers for CLR via thermodynamic screening, and select nanostructured carriers were synthesized and evaluated for their oxidation kinetics and high-temperature stability in CLR with pure CO2 and H2O, respectively, as well as mixed CO2 and H2O streams. The latter case results in the direct formation of syngas streams with tunable CO:H2 ratios. Finally, the feasibility of CLR in a periodically operated fixed-bed reactor was evaluated using a simplified pseudo-homogeneous reactor model.

See more of this Session: Chemical Looping Processes I
See more of this Group/Topical: Topical E: High Temperature Environmentally Sustainable Energy Processes (sessions joint with the Environmental Division)

# A dual fluidized bed system for chemical looping combustion of solid fuels

*Tobias Pröll and Hermann Hofbauer*
*Vienna University of Technology, Institute of Chemical Engineering,*
*Getreidemarkt 9/166, 1060 Vienna, Austria.*
*T. +43-1-58801-15901, F. +43-1-58801-15999, email: tobias.proell@tuwien.ac.at*

## Abstract

If it comes to direct use of solid fuels in chemical looping combustion (CLC) systems, three main challenges add to the general aspects valid for CLC of both gaseous and solid fuels:

- Conversion of volatile compounds formed in the upper part of the fuel reactor
- Conversion of char in the fuel reactor governed by slow gasification reactions
- Separation of fuel ash from oxygen carrier material

A fluidized bed reactor system design is proposed in which the solids flow in counter-current to the gas phase through a circulating fluidized bed fuel reactor. The bulk solids stream from the air reactor is added close to the top of the fuel reactor and leaves the fuel reactor at the bottom towards the air reactor. Only the air reactor entrainment is relevant for global oxygen carrier circulation. The division of the fuel reactor in consecutive zones is achieved by simple wedged ring-type flow obstacles reducing the riser cross section. The gas velocity in the fuel reactor is chosen such that solids entrainment is low. The solid fuel is fed in the middle of the fuel reactor. Due to the dispersed downward movement of the solids, volatiles are not produced in the upper part of the fuel reactor and the problem of insufficient gas phase conversion is avoided. Moreover, as a consequence of the series of mixed cells type residence time distribution of the solids, char conversion is expected to improve compared to well mixed fuel reactor inventories. It is likely that size classification effects are amplified by the proposed design, which may allow for selective ash removal. First cold flow model results show the fluid dynamic feasibility of the proposed concept. The system is expected to be a key to make CLC work directly for solid fuels.

## Introduction

Chemical looping combustion (CLC) is an innovative process for oxidizing fuels using metal oxides as oxygen carriers in order to transport oxygen from combustion air to fuel. Two separate exhaust gas streams are obtained and $CO_2$ occurs together with $H_2O$ from the fuel reactor while $N_2$ and excess $O_2$ are released from the air reactor (Fig. 1). The global heat release from CLC is equal to that of direct combustion, with most of the heat resulting from the strongly exothermic oxidation of oxygen carrier in the air reactor. Although proposed already about 60 years ago [1], CLC has recently gained interest as a technology for carbon capture from combustion processes [2-8]. CLC shows a unique potential because gas-gas separation and the related energy penalty are inherently avoided.

With respect to CLC technology development, there are generally two main problem areas: the selection, design and production of suitable oxygen carrier materials on the one hand and the design of suitable reactor configurations on the other hand.

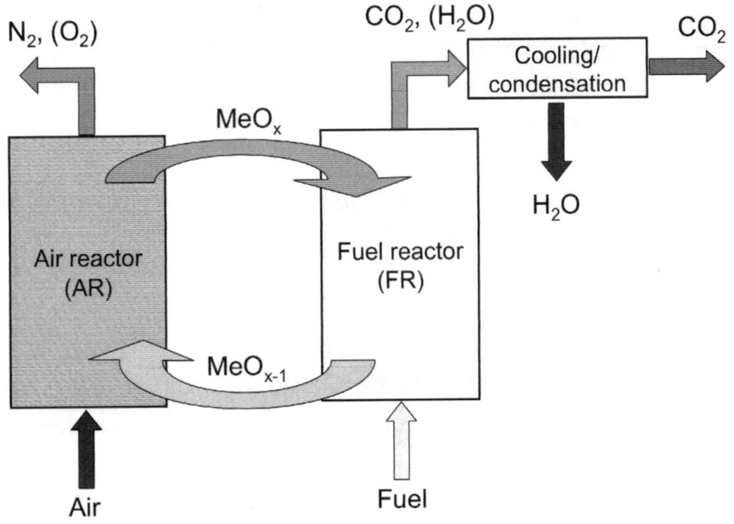

Fig. 1: Chemical looping combustion (CLC) for gaseous fuels.

For gaseous fuels, significant progress has been achieved recently leading to proof of concept with satisfactory fuel conversion performance at 140 kW in a dual fluidized bed system using oxygen carrier particles based on nickel [9]. Even though there is ongoing activity in further optimizing particles with respect to reactivity, mechanical strength and production costs, as well as with respect to scale up of fluidized bed reactor systems, it is widely agreed today that the technology will be technically feasible at large scale in the case of gaseous fuels. $CO_2$ ready industrial steam generation, for instance, can be done using CLC without any significant energy penalty [10].

If it comes to power production from gaseous fuels, the problem is that CLC works at atmospheric pressure and would just deliver heat for steam cycle operation. Maximum fuel to electricity efficiencies of about 46 % can be expected in this case. For gaseous fuels, however, the competing gas turbine combined cycles reach up to 58 % efficiency and may be competitive to CLC even if first generation $CO_2$ capture technologies such as oxyfuel firing are applied. Thus, we see two long-term possibilities to use CLC for power production: the direct use of solid fuels in combination with steam cycles and, on the other hand, pressurized operation of CLC to drive gas turbine combined cycle configurations.

Configurations with pressurized CLC and externally fired gas turbines have been theoretically studied [4, 11, 12] but it seems there are no promising solutions available regarding suitable CLC systems for pressurized operation. A special challenge is that relatively high solids circulation is required to provide enough oxygen and, on a density basis, this requirement increases proportional to pressure. With regard to the direct use of solid fuels at atmospheric pressure, the situation looks somewhat more encouraging. Given the tremendous potential in the area of clean coal to power technologies, a large number of research groups are actively working on direct coal CLC worldwide [8].

With respect to the choice of oxygen carrier material for solid fuel CLC, there are again two general possibilities. Firstly, one could aim at low-cost natural minerals typically based on iron oxides such as ilmenite ($FeTiO_3$) or waste material from steel production [13, 14]. These materials would justify a certain replacement rate in order to account for ash accumulation and attrition. Secondly, designed oxygen carriers with the thermodynamic potential to release free $O_2$ in the fuel reactor based on $CuO/Cu_2O$ or on $Mn_2O_3/Mn_3O_4$ could significantly enhance fuel conversion and would allow for full oxidation of volatiles. This is referred to as chemical looping with oxygen uncoupling (CLOU) [15].

The aim of the present contribution was to review the main requirements for direct solid fuel reactor systems and to develop a technically promising design proposal. The focus is on operation with "normal" oxygen carriers even though the findings might as well be relevant for CLOU and even for CLC of gaseous fuels.

## Theoretical background

The general principle of CLC for solid fuels is shown in Fig. 2.

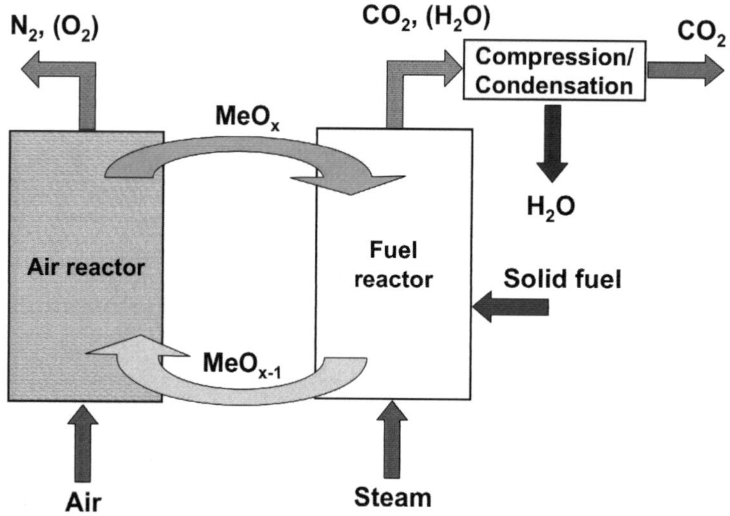


Fig. 2: Direct use of solid fuels in CLC with steam as gasification agent.

If solid fuels are fed into a steam-fluidized hot bed of oxygen carrier particles, the fuel particles undergo three consecutive steps of drying, devolatilization and char decomposition by gasification. The gaseous reaction products are then available for oxidation in contact with the oxygen carrier particles. It is evident that sufficient contact to the oxygen carrier must be guaranteed for the gaseous decomposition products in order to achieve high gas phase conversion to $CO_2$ and $H_2O$. However, it was found that decomposing particles tend to segregate at the bed surface in case of bubbling fluidized beds [16]. Even in the case of circulating fluidized bed (CFB) fuel reactors with uniform fuel distribution, some volatiles are formed close to the gas-solid separator and thus not have time to be fully oxidized. As a consequence, a suitable fuel reactor design must guarantee oxygen carrier contact of the gas stream leaving the devolatilization zone.

A second major problem is the slow gasification reaction governing the conversion of the remaining char left after devolatilization of the fuel. These char particles are expected to mix well with the solids and the residence time distribution of the bulk of solids can also be considered to behave well mixed in fluidized bed reactors. Therefore, a part of the char will always be lost to the air reactor in single stage fluidized bed fuel reactors. As a consequence, a suitable fuel reactor design will feature a long residence time for solids and, preferably, a residence time distribution that prevents direct char slip to the air reactor. This need has already been accounted for in the design of a 10 kW continuous unit [17]. Alternatively, methods to selectively separate char particles from the oxygen carrier leaving the fuel reactor and to recycle the char into the fuel reactor have been proposed [18].

Finally, solid fuels contain ash that may accumulate in the bed of oxygen carrier particles. Selective removal of ash particles is desired in order to minimize oxygen carrier renewal rates. One possible principle here is pneumatic separation.

The first two requirements would be fulfilled for a reactor system according to Fig. 3. As early as in 1951 Lewis and Gilliland [1] were aware of the need for a fuel reactor design organized in stages where gas and solids are directed in counter-current flow through discrete bubbling bed zones. However, such systems look complex to be built at large scale and, indeed, none of the known recent practical efforts follows this approach [8].

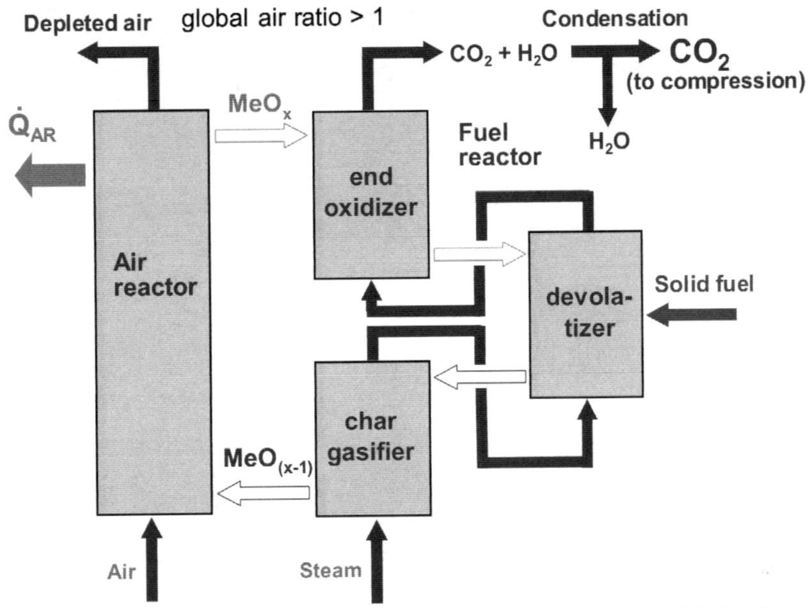


Fig. 3: Theoretical configuration to achieve volatiles oxidation and high char conversion rates.

The relevant task can be formulated, therefore, as to find a most simple solution to achieve the main functionalities of the theoretical system in Fig. 3 and, in addition, selectively separate fuel ash from the solids loop.

## Multistage circulating fluidized bed riser approach

The starting point for the presented design proposal was the dual circulating fluidized bed (DCFB) system successfully applied to CLC of gaseous fuels [19]. In this system, the fuel reactor is designed as CFB but the solids entrainment of the fuel reactor is not relevant for global solids circulation because the solids return to the air reactor through a loop seal connected to the bottom of the fuel reactor. Bu and Zhu [20] proposed and investigated the use of ring-type flow obstacles along the height of a CFB riser to intensify gas-solids contact by repeatedly disturbing the formation of a core-annulus profile. They show that it is possible to increase the solids density globally and that the solids density is high in the low velocity areas between the obstacles and low in the high velocity areas of reduced cross section. If this approach is combined with the DCFB concept, where bed material from the air reactor can be continuously added close to the top of the fuel reactor, a dispersed counter-current flow of solids and gas can be obtained if the fuel reactor is fluidized such that final solids entrainment is low. Additionally, it can be expected that fine ash particles are concentrated in the small solids stream entrained from the staged fuel reactor. Thus, selective ash separation is possible from the fuel reactor's internal solids return loop. The proposed system configuration is sketched in Fig. 4.

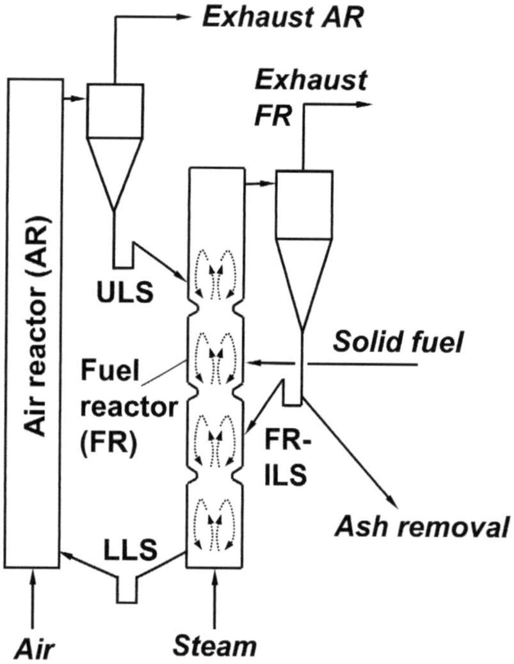


Fig. 4: Proposed dual circulating fluidized bed design with dispersed gas-solid counter-current in the fuel reactor (Steam fluidized loop seals: LLS...Lower loop seal, ULS...Upper loop seal, FR-ILS...Fuel reactor internal loop seal).

The following general observations can be made for the proposed system:

- The fluidization gas velocity has likely to be in bubbling or turbulent regime in the free cross section and in turbulent to fast regime in the reduced cross section. Specific design and fluidization criteria need to be experimentally assessed.
- The pressure drop of the fuel reactor is expected to be significantly increased due to increased solids inventory. This needs to be accounted for when designing the lower loop seal and the relative elevation of the air reactor bottom.
- The bulk of particles inside each zone is lifted by the fast gas stream in the reduced cross section and freely falls back as the gas velocity decreases in the open cross section areas. Therefore, significant acceleration phenomena are expected and the local pressure profile cannot be directly related to solids concentration in this case.
- For inventories with a certain size and/or density distribution it can be expected that larger or denser particles move downward through the fuel reactor more quickly than smaller or lighter particles.

## Cold flow model study

In order to study the fluid dynamics in more detail, an existing cold flow model of a DCFB pilot plant [21] was equipped with three flow obstacles and operated with air and bronze powder in order to meet the scaling criteria. Air reactor fluidization conditions and total bed inventory were kept constant. Thus, the solids circulation rate between air reactor and fuel reactor was approximately constant, too. The fuel reactor fluidization rate was varied in order to see the effect of gas velocity on solids distribution in the fuel reactor. The fuel reactor internal loop seal fluidization rate needed to be adjusted according to the amount of solids entrained from the fuel reactor. The results are shown together with a sketch of the cold flow model in Fig. 5. For the lowest fuel reactor fluidization rate of 3.4 $Nm^3/h$ the solids are concentrated in the lower part of the fuel reactor. At the medium fluidization rate of 5.0 $Nm^3/h$ more solids are present in the upper part and slight solids entrainment is achieved from the fuel reactor. It is important to notice that the highest pressure drop occurs across the obstacles although the solids concentration is visually lower in these areas. This effect was also observed by Bu and Zhu [20] and can be attributed to particle acceleration phenomena in the obstacle region. The particles take up kinetic energy in the obstacle region, what results in an increased pressure drop. In the low gas velocity sections between the obstacles, particles fall freely back, mainly along the wall, are not lifted by the gas stream and therefore cause little pressure drop here. At the highest fluidization rate of 6.5 $Nm^3/h$, the gas flow is strong enough to prevent the particles from moving downward through the obstacle regions. Significant solids entrainment is observed from the fuel reactor. The maximum load of the counter-current column is reached.

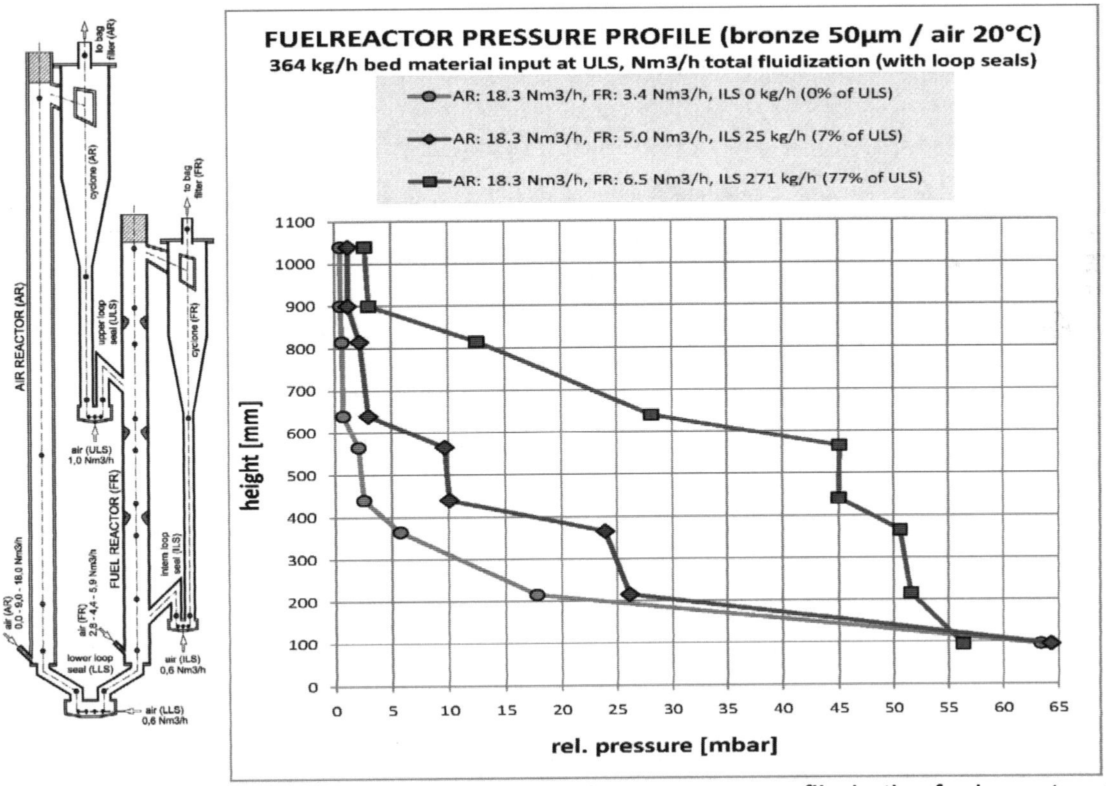


Fig. 5: Effect of fuel reactor fluidization rate on the pressure profile in the fuel reactor. Cold flow model sketch with obstacles and pressure measurement points indicated in the corresponding heights.

## Conclusion

Combining the previously proposed dual circulating fluidized bed (DCFB) reactor system with the idea of dividing the fuel reactor in vertical sections by the use of wedged rings as flow obstacles makes it possible to achieve a dispersed counter-current flow pattern of gas phase and solids in the fuel reactor. Thus, a relatively small geometrical change may be a key to make dual bed systems significantly more efficient in fuel conversion. In the case of CLC for solid fuels, the proposed design may be a key to achieve satisfactory conversion of both char and volatiles. It is expected that implementation of the design in refractory-lined reactors is not a major issue. Since the gas velocities need to be chosen lower than in comparable CFB risers without obstacles, erosion of refractory can be expected to stay within tolerable limits.

Further research is definitely required in order to determine the necessary flow characteristics and design rules for such systems and to achieve proof of concept at a relevant scale.

## Acknowledgement

Valuable support by Diana Carolina Guio-Perez (literature review on CFB internals, cold flow model preparation), Johannes Schmid (cold flow model operation and preparation of Fig. 5) and Klemens Marx (design and procurement of flow obstacles) is gratefully acknowledged.

## References

[1] Lewis, W.K., Gilliland, E.R., 1954, "Production of pure carbon dioxide", United States Patent No. 2,665,972 (filed 1951).

[2] Ishida, M., Jin, H., 1997, "CO2 recovery in a power plant with chemical looping combustion", Energy Conversion and Management, 38(Suppl. 1), 187–192.

[3] Lyngfelt, A., Leckner, B., Mattisson, T., 2001, "A fluidized bed combustion process with inherent CO2 separation, Application of Chemical Looping Combustion", Chemical Engineering Science, 56, 3101–3113.

[4] Wolf, J., Anheden, M., Yan, J., 2001, "Performance Analysis of Combined Cycles with Chemical Looping Combustion for CO2 Capture", Proceedings of 18th Pittsburg Coal Conference, December 3–7, Newcastle, NSW, Australia, Session 23.

[5] Ryu, H.-J., Bae, D.-H., Jin, G.-T., 2002, "Chemical-looping combustion process with inherent CO2 separation; Reaction kinetics of oxygen carrier particles and 50kWth reactor design", in: The World Congress of Korean and Korean Ethnic Scientists and Engineers, Seoul, Korea, pp.738–743.

[6] Hossain, M.M., de Lasa, H.I., 2008, "Chemical-looping combustion (CLC) for inherent CO2 separations – a review", Chemical Engineering Science, 63(18), 4433–4451.

[7] Lyngfelt, A., Johansson, M., Mattisson, T., 2008, "Chemical-looping combustion – status of development", in: Proceedings 9th International Conference on Circulating Fluidized Beds (CFB-9), May 13-16, 2008, Hamburg, Germany.

[8] Fang, H., Haibin, L., Zengli, Z., 2009, "Advancements in development of chemical-looping combustion: A review", International Journal of Chemical Engineering, art. no. 710515.

[9] Kolbitsch, P., Bolhàr-Nordenkampf, J., Pröll, T, Hofbauer, H, 2009, "Comparison of two Ni-based oxygen carriers for chemical looping combustion of natural gas in 140kW continuous looping operation", Industrial and Engineering Chemistry Research, 48(11), 5542–5547.

[10] Nikoo, M., 2010, "Application of chemical looping combustion for bitumen extraction via Steam Assisted Gravity Drainage (SAGD) in Alberta's oil sands", 2nd Network Meeting High Temperature Solid Looping Cycles, September 15-17, Alkmaar, NL, available online at: www.ieaghg.org.

[11] Brandvoll, Ø., Bolland, O., 2004, "Inherent CO2 capture using chemical looping combustion in a natural gas fired power cycle", Journal of Engineering for Gas Turbines and Power, 126(2), 316-321.

[12] Naqvi, R., Bolland, O., 2007, "Multi-stage chemical looping combustion (CLC) for combined cycles with CO2 capture", International Journal of Greenhouse Gas Control, 1(1), 19-30.

[13] Leion, H., Lyngfelt, A., Johansson, M., Jerndal, E., Mattisson, T., 2008, "The use of ilmenite as an oxygen carrier in chemical-looping combustion", Chemical Engineering Research and Design, 86(9), 1017-1026.

[14] Leion, H., Mattisson, T., Lyngfelt, A., 2009, "Use of ores and industrial products as oxygen carriers in chemical-looping combustion", Energy and Fuels, 23(4), 2307-2315.

[15] Mattisson, T., Lyngfelt, A., Leion, H., 2009, "Chemical-looping with oxygen uncoupling for combustion of solid fuels", International Journal of Greenhouse Gas Control, 3(1), 11-19.

[16] Fiorentino, M., Marzocchella, A., Salatino, P., 1997, "Segregation of fuel particles and volatile matter during devolatilization in a fluidized bed reactor – Part 2: Experimental", Chemical Engineering Science, 52, 1909-1922.

[17] Berguerand, N., Lyngfelt, A., 2008, "Design and operation of a 10 kWth chemical-looping combustor for solid fuels - Testing with South African coal", Fuel 87(12), 2713-2726.

[18] Berguerand, N., Lyngfelt, A., 2008, "The use of petroleum coke as fuel in a 10 kWth chemical-looping combustor", International Journal of Greenhouse Gas Control, 2(2), 169-179.

[19] Pröll, T., Kolbitsch, P., Bolhàr-Nordenkampf, J., Hofbauer, H., 2009, "A novel dual circulating fluidized bed system for chemical looping processes", AIChE Journal, 55(12), 3255-3266.

[20] Bu, J., Zhu, J.-X., 1999, "Influence of ring-type internals on axial pressure distribution in circulating fluidized bed", Canadian Journal of Chemical Engineering, 77(1), 26-34.

[21] Pröll, T., Rupanovits, K., Kolbitsch, P., Bolhàr-Nordenkampf, J., Hofbauer, H., 2009, "Cold flow model study on a dual circulating fluidized bed system for chemical looping processes", Chemical Engineering and Technology, 32(3), 418-424.

# (321f) Calcium Looping Process (CLP) for Enhanced Hydrocarbon Reforming

**Nihar Phalak**, Shwetha Ramkumar, Niranjani Deshpande and Liang-Shih Fan, William G. Lowrie
Department of Chemical and Biomolecular Engineering, The Ohio State University, Columbus, OH

Increasing energy needs have encouraged the development of alternate sources of energy, but fossil fuels are likely to continue to play an important role in satisfying this demand in the near future. The Calcium Looping Process (CLP) is being developed at The Ohio State University (OSU) to enable clean conversion of hydrocarbons to hydrogen, which can have potential applications in electricity generation, liquid fuels production or chemical synthesis.

The presence of a calcium-based sorbent along with the traditional reforming catalyst minimizes the equilibrium limitation of the reforming and water gas shift (WGS) reactions enabling the production of high-purity H2 with in-situ CO2 capture and sulfur removal in a single step at high temperatures. The CO2 capture by the sorbent enhances the reforming and WGS reactions thus improving the yield of H2. The integration of several unit operations in a single reactor reduces the overall process foot print of the hydrogen production process.

The CLP process has been studied in a laboratory-scale setup and various parameters like temperature, pressure and steam requirements have been investigated with respect to hydrogen purity and impurities removal. High temperature and high pressure hydration of the calcium sorbent to maintain its reactivity over multiple cycles has also been investigated and the effect of different calcination (sorbent regeneration) conditions on multicyclic operation has been tested. These tests have shown that high purity hydrogen production is possible at lower temperatures and lower steam to carbon ratios as compared to conventional reforming reaction. Preliminary attempts at using sorbent hydration for multicyclic runs have yielded encouraging results.

See more of this Session: Chemical Looping Processes I
See more of this Group/Topical: Topical E: High Temperature Environmentally Sustainable Energy Processes (sessions joint with the Environmental Division)

# (391a) Copper as An Oxygen Carrier in a Chemical Looping Combustion System: Reaction Kinetics and Fluidized Bed Performance

**Christopher K. Clayton**, Chemical Engineering, University of Utah, Salt Lake City, UT and Kevin Whitty, Institute for Clean and Secure Energy, The University of Utah, Salt Lake City, UT

In order to better classify potential oxygen carriers for a Chemical-Looping Combustion system, a lab-scale bubbling fluidized bed reactor has been designed and built at the University of Utah. In order to simulate a dual-fluidized bed system, the reacting gas is cycled between the oxidizing gas (Air) and the fuel gas (CH4 or H2/CO mixture). Volume percentages of CH4, CO, CO2 and 02 are analyzed in real time by a CAI NDIR/O2 gas analyzer. Copper has been identified as a potential CLOU (chemical looping with oxygen uncoupling) material. A CLOU material has the ability to both bind and release oxygen under different atmospheres without the aid of a reducing agent. This is particularly useful for combustion of solid fuels. In order to overcome the low melting point and associated sintering propensity of pure copper, a titanium oxide support was used. Tests to characterize performance of the supported copper as an oxygen carrier have been conducted at temperatures ranging from 750°C to 950°C. One test was conducted for 24 hours in order to determine the long term performance of copper. Development of internal surface area and particle morphology as a function of the number of looping cycles is presented. Such data is important when considering applicability in industrial systems where attrition resistance and long-term reactivity are important. This paper describes the experimental approach, data analysis and assessment of the potential for copper as a carrier in full-scale systems.

See more of this Session: Chemical Looping Processes II
See more of this Group/Topical: Topical E: High Temperature Environmentally Sustainable Energy Processes (sessions joint with the Environmental Division)

# Operation of a New 10kW CLC Unit with Independant Solid Flow Control

A.Hoteit, IFP
K. Surla, IFP
M.M.Yazdanpannah, IFP
Sébastien Rifflart, Total
Thierry Gauthier, IFP

AIChE Annual Meeting,
Nov 7-12, 2010, Salt Lake City, Utah

Chemical Looping: Motivation and Technology concept

Oxygen carriers developement

Cold flow model results

First experimental results on 10 kW pilote

Conclusions

TOTAL



IFP Energies nouvelles

## Chemical Looping Combustion
## General motivation

- Develop the potential breakthrough $CO_2$ capture technology for power generation
  - No cryogenic Air Separation Unit required,
  - In-situ $CO_2$ separation.
- Investigate the extension of CL processes: $CO_2$ free syngas and hydrogen production:
  - Chemical looping as base technology for reforming/gasification,
  - Fuel flexible concept (natural gas, refinery gas, oil, oil residues, coal, Petcoke and biomass).

3 TOTAL

IFP Energies nouvelles

# Chemical Looping development
# Key Global Team researches

## packed-bed :

- TNO

## Fluidized bed

Gaseous Fuel:

- Chalmers: **Prototype 10 kWt & 300 Wfuel**
- CSIC: **Prototype 5 & 10 kWt**
- Korea Institute of Energy Research
  **Prototype 50 kWt**
- Vienna University of Technology
  **120 $kW_{th}$ Gaseous Fuel CLC**
- IFP-Total
  **10 $kW_{th}$**

Solid Fuel:

- Chalmers (10 kW)
- Alstom
- Korea Institute of Energy Research

Liquid Fuel:

- IFP-Total

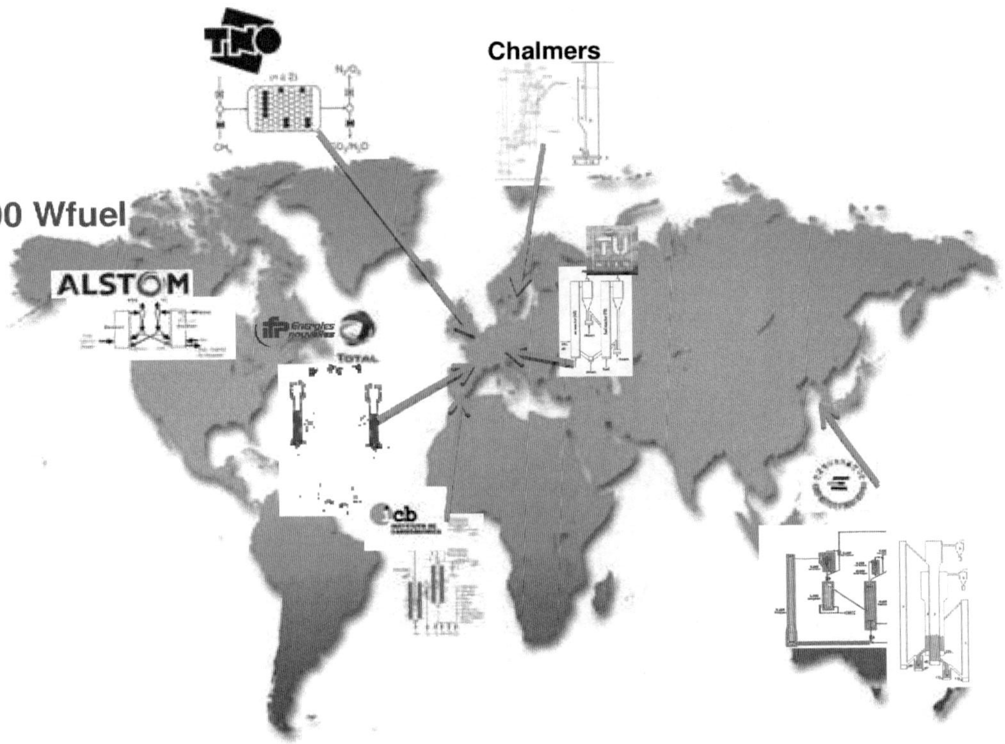


4 TOTAL

# Chemical Looping Combustion – What is it?

- **Use a solid to transport oxygen**
- **Two (at least) reactors**
  - Air reactor – oxidation of the solid
  - Fuel reactor – Burning the fuel with the oxygen from the solid
- **Can get a stream of almost pure $CO_2$, free of nitrogen**

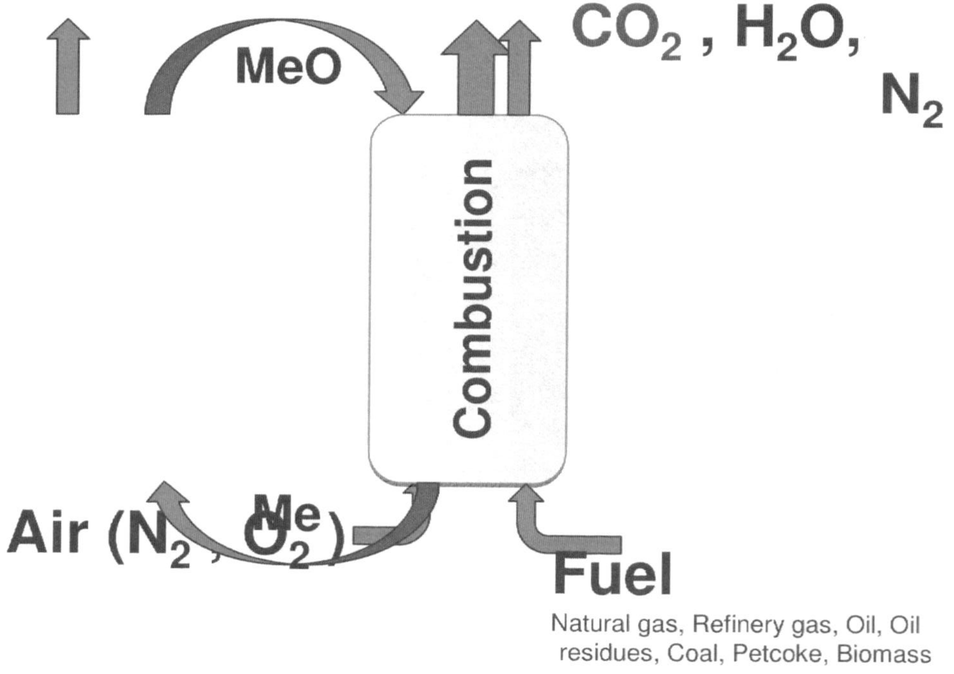


$$C_nH_{2m} + (2n+m)M_yO_x \rightarrow nCO_2 + mH_2O + (2n+m)M_yO_{x-1}$$ [Endothermic, 950° C]

$$M_yO_{x-1} + \frac{1}{2}O_2 \rightarrow M_yO_x$$ [Exothermic, 1030° C]

5 TOTAL

# Chemical Looping Combustion Basic principle & Background

- **Why do it?:** **Potentially lowest Cost Option for Capturing $CO_2$**
  **Competitive with and without $CO_2$ capture**

- **What is it?:** **Oxy-Firing without Oxygen Plant**

  - Solid Oxygen Carrier Circulates between Oxidizer and Reducer
  - Oxygen Carrier: Carries Oxygen and Heat
  - Carrier picks up $O_2$ in the Oxidizer, leaves $N_2$ behind
  - Carrier Burns the Fuel in the Reducer
  - Heat produces Steam for Power

**Oxygen Carriers:**

- **Metal Oxides:** Fe, Ni, Cu…Ores or on Substrates
- **Limestone-based** carriers

6 TOTAL

IFP Energies nouvelles

Chemical Looping: Motivation and Technology concept

Oxygen carriers developement

Cold flow model results

First experimental results on 10 kW pilote

Conclusions

# Chemical Looping Combustion
# Technical highlights: MeOx development route

**Qualification of oxide material: downselect $O_2$ carrier for pilot scale tests.**

- 80 options tested,
- Optimum oxide carrier selected,
- Same metal oxide particles used in 300 h tests

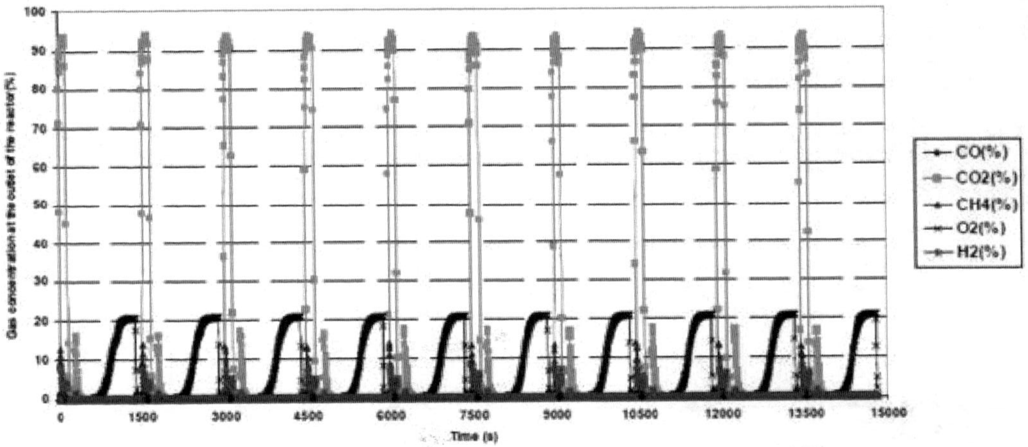


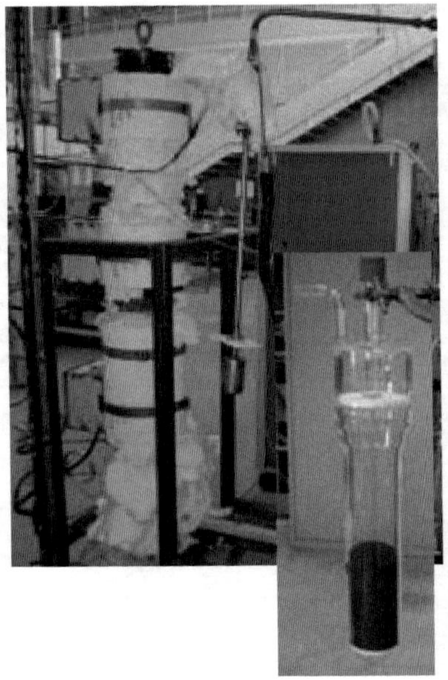

TOTAL

IFP Energies nouvelles

- Chemical Looping: Motivation and Technology concept
- Oxygen carriers developement
- Cold flow model results
- First experimental results on 10 kW pilote
- Conclusions

TOTAL

ifP Energies nouvelles

# Chemical Looping Combustion
# Cold flow model results

**High temperature limits the use of mechanical valves for CLC applications**

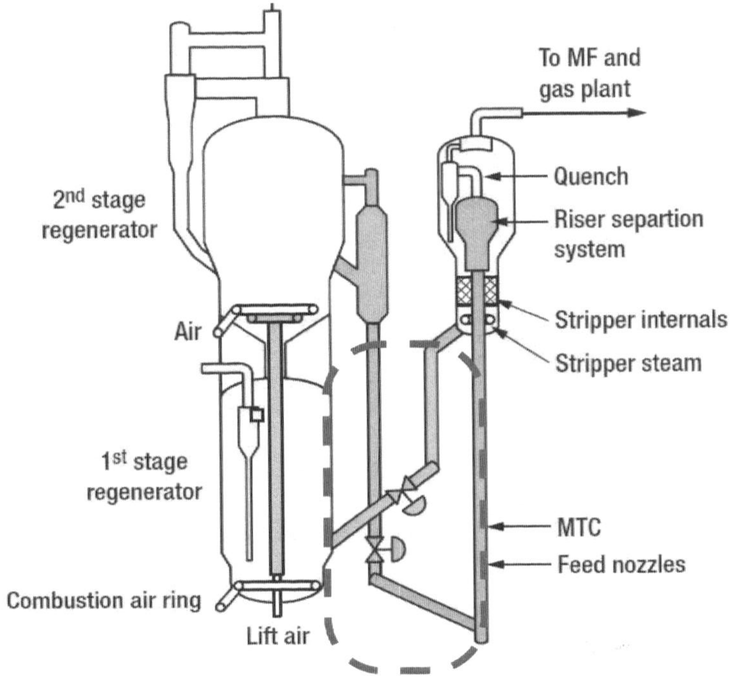


**Mechanical Valve**

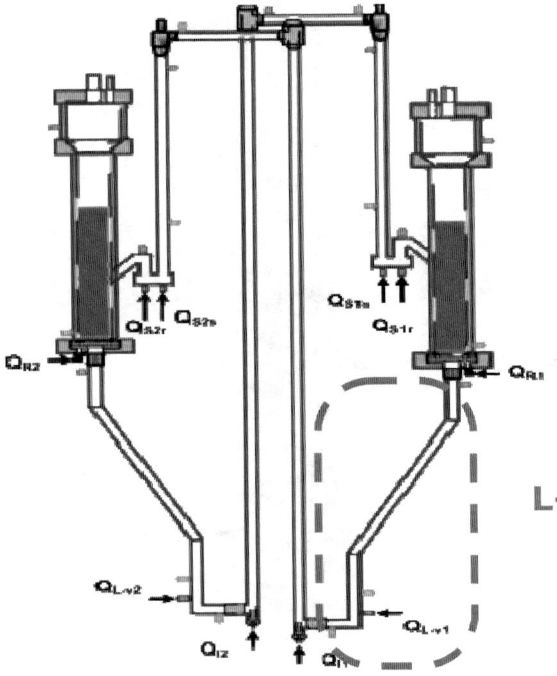

**L-valve**

# Chemical Looping Combustion
# Cold flow model results

- Design validation
- Solid flow control
- Correlation development
- Pressure drop and solid circulation modelling

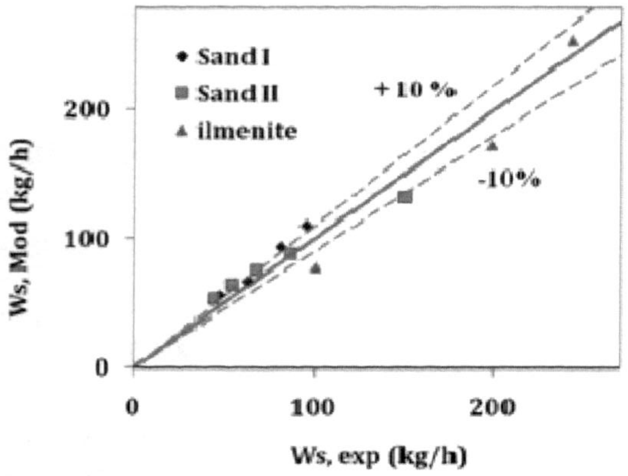


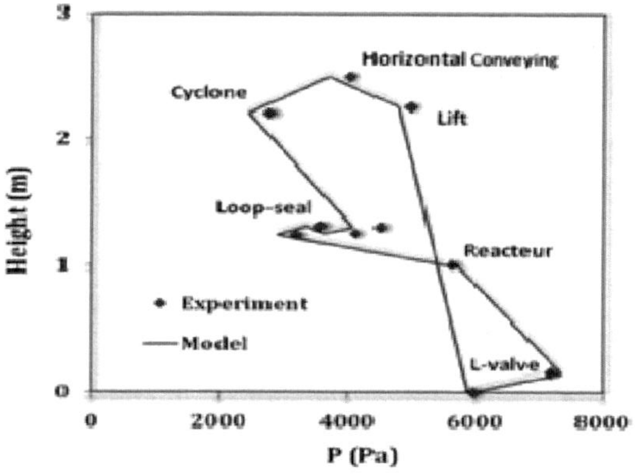


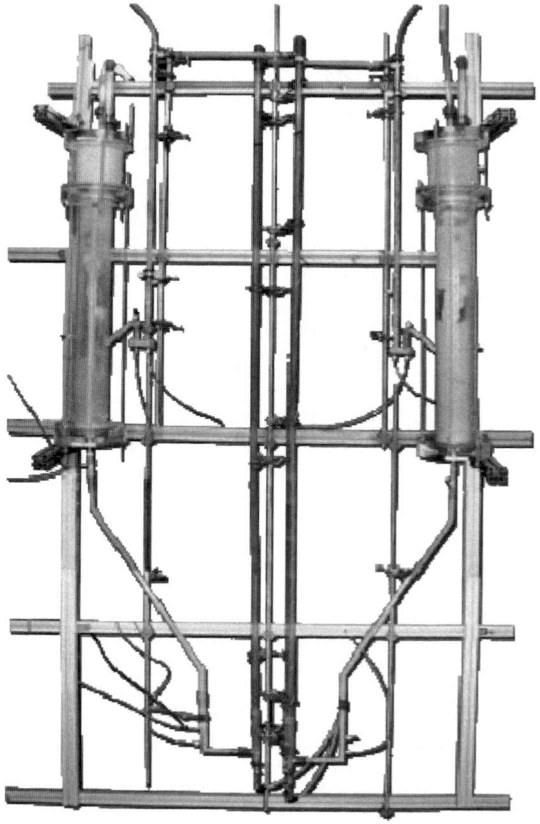



Energies nouvelles

Chemical Looping: Motivation and Technology concept

Oxygen carriers developement

Cold flow model results

First experimental results on 10 kW pilote

Conclusions

# Chemical Looping Combustion
## First experimental results on 10 kW pilote

- **Oxygen carrier used $NiO/NiAl_2O_4$ (60/40)**
- **Particle size fraction (100 – 300 µm)**
- **Experimental conditions used:**

**$CH_4$=0.2 Nm3/h + $N_2$=0.3 Nm$^3$/h**
**$CH_4$=0.4 Nm3/h + $N_2$=0.1 Nm$^3$/h**
**$CH_4$=0.5 Nm3/h**
**$CH_4$=0.75 Nm3/h**
**Temperature = 750-800-850°C**

TOTAL

ifp Energies nouvelles

# Chemical Looping Combustion
# First experimental results on 10 kW pilote

## Solid flow rate

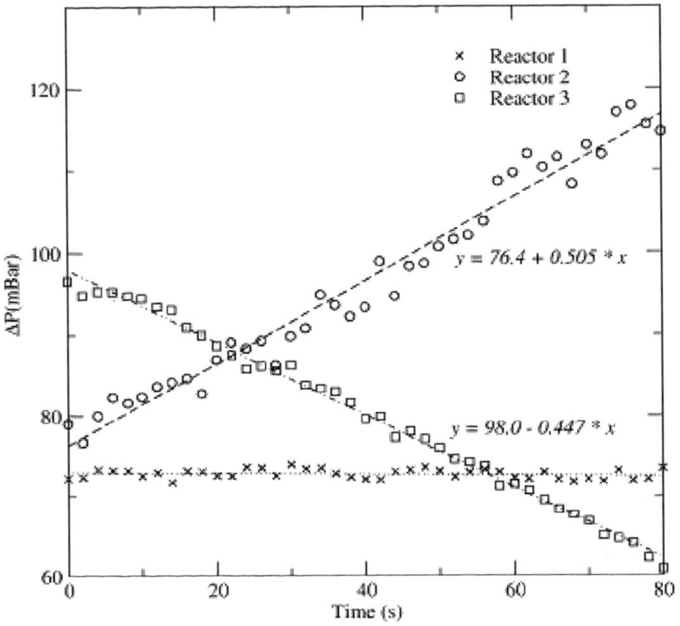


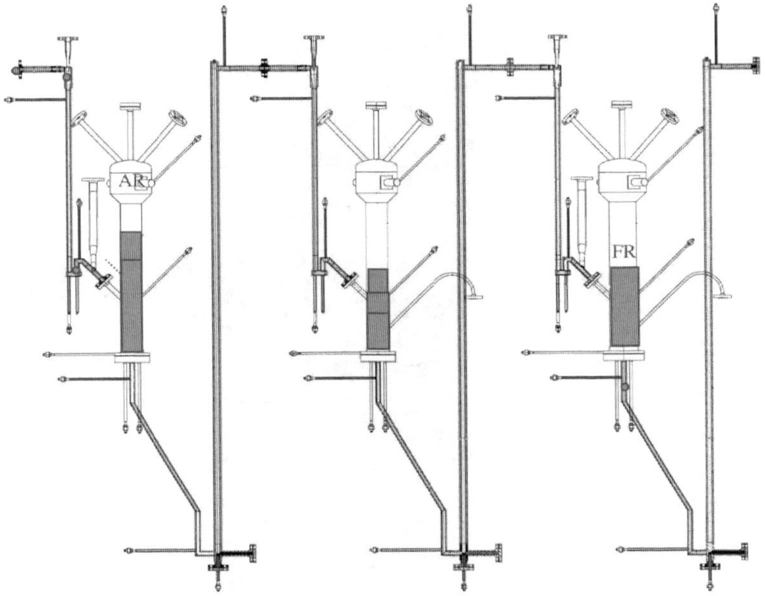


$$W_s = \frac{dM_{s,R}}{dt} = \frac{A_R}{g} \cdot \frac{dP_R}{dt}$$

# Chemical Looping Combustion
# First experimental results on 10 kW pilote

## Solid circulation control and Stabilisation

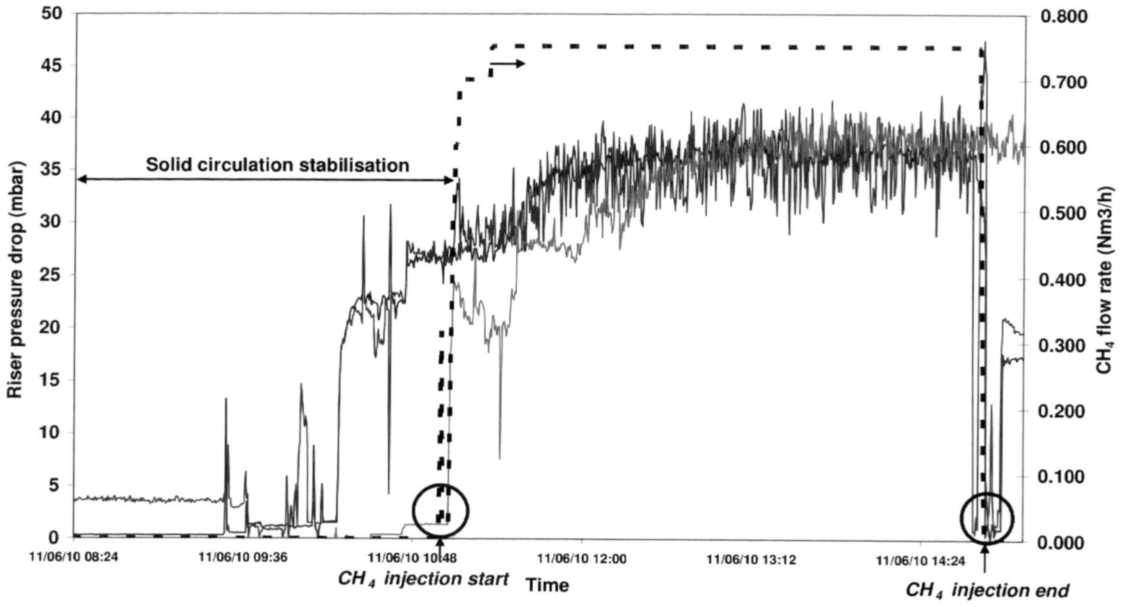


Solid flow rate : 135 kg/h

# Chemical Looping Combustion
# First experimental results on 10 kW pilote

## Reduction

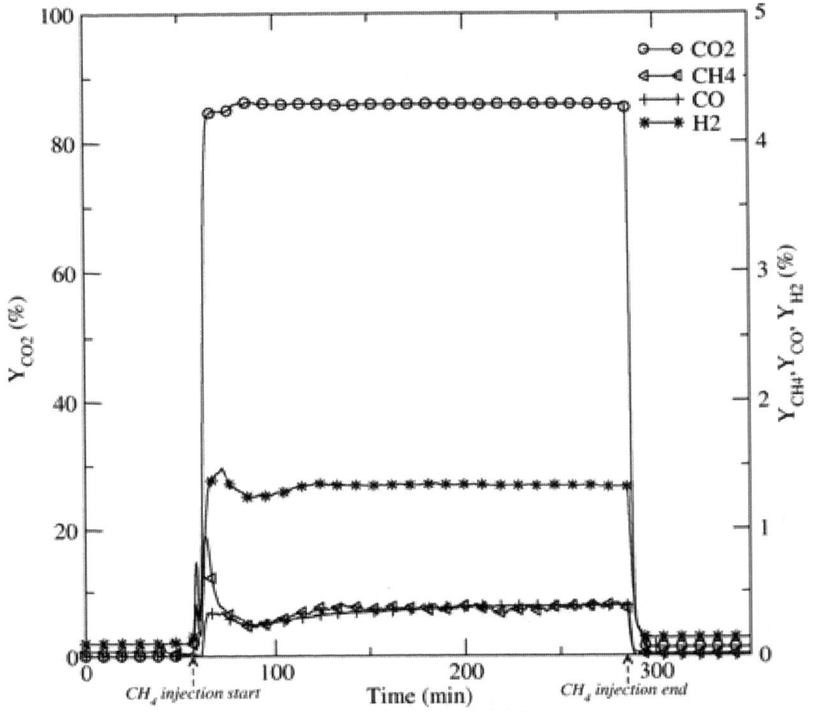


gas yield, $\gamma_{red}$,

$$\gamma_{red} = \frac{Y_{CO_2}}{Y_{CO_2} + Y_{CH_4} + Y_{CO}}$$

99.15% at 800° C

97.7% $CO_2$
1.5% $H_2$
0.4%CO
0.4% $CH_4$

**This composition is obtained after substituting of the nitrogen dilution coming from the loop-seal.**

TOTAL

ifp Energies nouvelles

# Chemical Looping Combustion First experimental results on 10 kW pilote

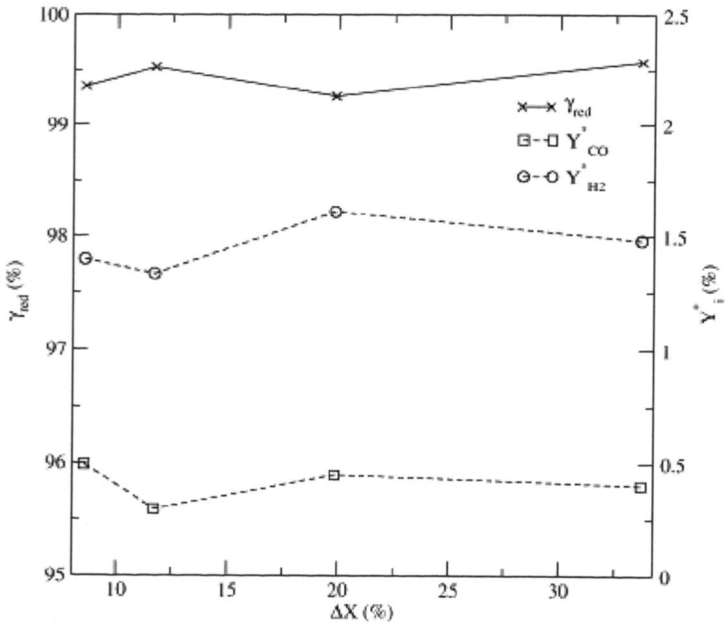


| Methane flow inlet (Nm3/h) | solid circulations flow rate (kg/h) | ΔX (%) |
|---|---|---|
| 0.5 | 33 | 33.7 |
| 0.5 | 56 | 19.9 |
| 0.5 | 95 | 11.0 |
| 0.5 | 130 | 8.6 |

$\gamma_{red} > 99\%$

1.5% $H_2$

0.4% CO

**Large flexibility of the process using NiO particles**

TOTAL

IFP Energies nouvelles

# Chemical Looping Combustion
# First experimental results on 10 kW pilote

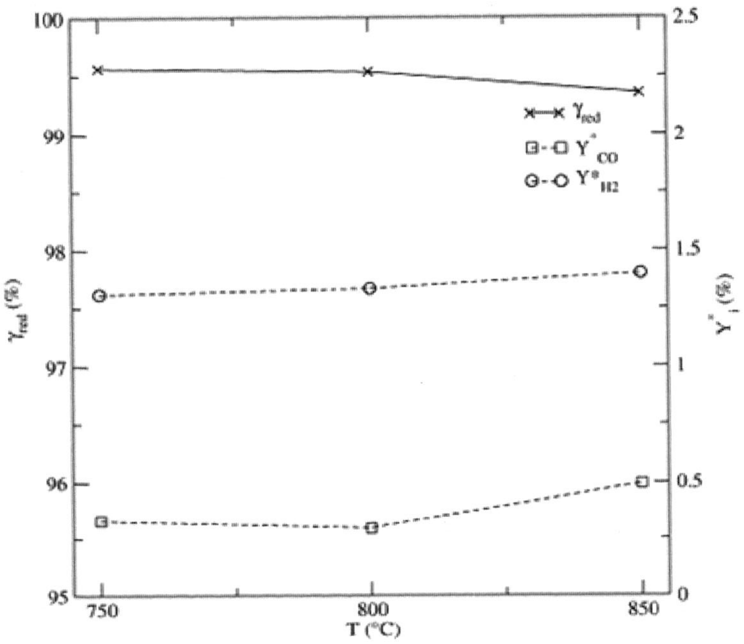


| Temperature (°C) | solid circulations flow rate (kg/h) |
|---|---|
| 750 | 56 |
| 800 | 56 |
| 850 | 56 |

$\gamma_{red} > 99\%$

**Thermodynamic limitation of NiO-Ni system at higher temperatures**

TOTAL

ifp Energies nouvelles

# Chemical Looping Combustion
# First experimental results on 10 kW pilote

## Oxydation

outlet R3= 2.6% $O_2$

outlet R2= 0% $O_2$

global air ratio of 1.05

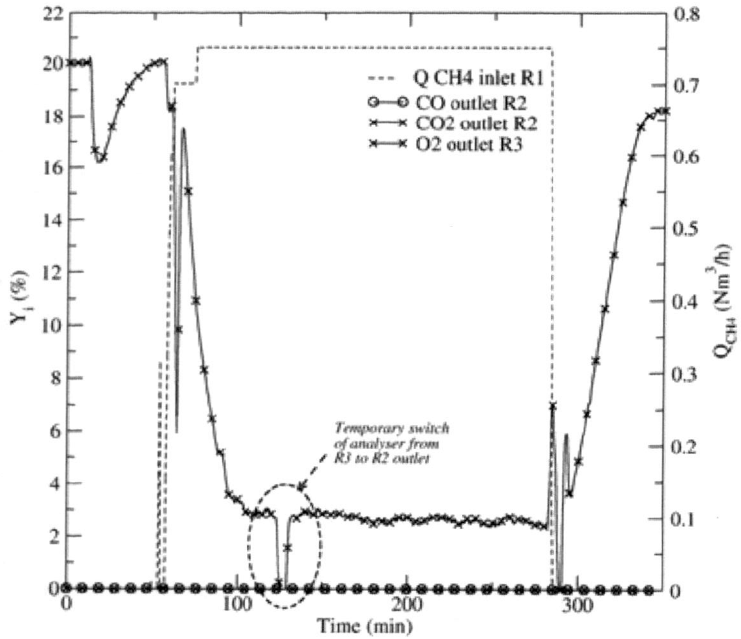


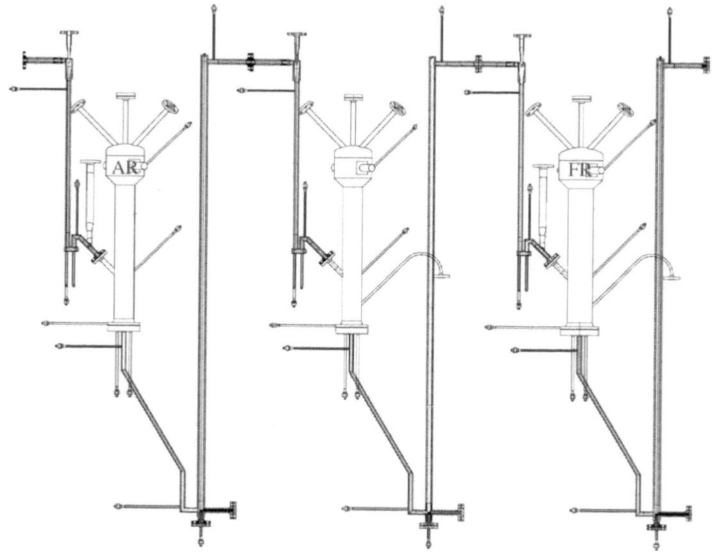


no CO or $CO_2 \rightarrow$ no coke is formed during reduction

100% of the $CO_2$ produced is captured

- Chemical Looping: Motivation and Technology concept
- Oxygen carriers developement
- Cold flow model results
- First experimental results on 10 kW pilote
- Conclusions

# Chemical Looping Combustion Conclusions

- Pilot design validation
- 10 kW test unit / NiO-based oxygen carrier:
  - no detectable carbon leakage to the air reactor: 100% capture
  - $CH_4$ conversion up to 99%
  - $CO_2$ yield up to 97% based on total carbon in fuel
  - high fuel conversion at air ratio 1.05
- Further work:
  - Other oxygen carrier materials tests
  - Different feed stocks liquid and/or Coal

21 TOTAL

IFP Energies nouvelles

# Operation of a New 10kW CLC Unit with Independant Solid Flow Control

A.Hoteit, IFP
K. Surla, IFP
M.M.Yazdanpannah, IFP
Sébastien Rifflart, Total
Thierry Gauthier, IFP

AIChE Annual Meeting,
Nov 7-12, 2010, Salt Lake City, Utah

# (391c) Syngas Chemical Looping - 25kWth Sub-Pilot Scale Testing

**Deepak Sridhar**, Andrew Tong, Fanxing Li, Ray Kim, Liang Zeng, Fei Wang and L.-S. Fan, William. G. Lowrie Dept. of Chemical and Biomolecular Engineering, The Ohio State University, Columbus, OH

Pursuit of economically attractive fossil fuel conversion technologies that produce clean energy carriers, in combination with carbon capture, is one of the main focuses of the ongoing energy research in the United States. The Syngas Chemical Looping (SCL) process, developed at The Ohio State University, is a promising solution that provides for 100% carbon capture from various carbonaceous fuels, while co-generating Hydrogen and electricity. A specially tailored Iron based composite medium is moved through three reactors namely; the reducer, the oxidizer and the combustor, to achieve the stated benefits. In the moving bed reducer, high concentration $CO_2$ stream is generated by fuel combustion using the oxygen provided by the looping medium. The moving bed oxidizer partly replenishes the depleted oxygen in the looping medium with steam, producing high purity $H_2$. The oxygen carrying capacity of the looping medium is completely restored and the loop is completed, in the combustor, which uses air to combust and entrain the looping medium while producing heat.

Over the past 6 years, the SCL process has been developed from a lab scale unit to a $25kW_{th}$ sub-pilot scale unit. Tests prior to the sub-pilot scale construction involved extensive research to find a working oxygen carrier looping medium in a lab scale and validating the operation of the individual reactors using a bench scale unit. The current article will focus on the operational results obtained from the $25kW_{th}$ sub-pilot SCL system. This unit has been operated for more than 40 hours and successful results have been obtained. The possibility to continuously generate high purity hydrogen and high concentration $CO_2$ stream in separate reactors has been confirmed. Post experiment looping medium analysis results will be discussed which include attrition results, reactivity testing and re-pelletization studies. The techno-economic performances of the SCL process as well as the design philosophy of a $250kW_{th}$ pilot scale unit are also discussed.

See more of this Session: Chemical Looping Processes II
See more of this Group/Topical: Topical E: High Temperature Environmentally Sustainable Energy Processes (sessions joint with the Environmental Division)

# (391d) Design of a Chemical-Looping Combustion System for Solid Fuels

Suman K. Hajra[1], **Thomas O'Brien**[1] and Kartikeya Mahalatkar[2], (1)Simulation, Analysis & Computational, NETL/DOE, Morgantown, WV, (2)ANSYS-Fluent, Inc., Morgantown, WV

Fossil fuels, such as coal, oil and natural gas are widely used in the power and industrial sectors. Their demand is gradually increasing in spite of a growing international concern about climate change and an increasing desire to control greenhouse gas (GHG) emissions. Carbon capture and storage (CCS) has significant potential to decrease anthropogenic carbon dioxide emissions. A great majority of the carbon dioxide capture R&D is focused on the power generation sector, particularly coal-fired power generation. Chemical looping combustion (CLC) promises to by an energy efficient carbon capture technology because of its inherent separation of carbon dioxide. A CLC system usually consisted of two fluidized bed reactors, an air reactor (AR) and a fuel reactor (FR), and uses a granular oxygen carrier to transfer oxygen from combustion air to the fuel. The oxygen carrier is oxidized by the air in the AR and is reduced by the fuel in the FR. Hence, air is not mixed with the fuel, and the outgoing carbon dioxide stream is not diluted by nitrogen. Successful commercialization of power generation processes using coal directly, utilizing CL technology, requires a suitable process design, which is currently under development. A steady state model of a CLC process has been developed, based on mass and energy balance. The model is used to estimate the reactors dimensions using iron oxide as oxygen carrier, as an example. Coal is represented in a general way so as to be able to include a wide range of fuels. Mass balance is based on an effective molecular formula for the carrier particles and the solid circulation rate is partitioned into metal, oxygen and inert flow rates. The solid circulation rate in the system is calculated as a function of the difference in conversion of the carrier particles and other operating parameters.

See more of this Session: Chemical Looping Processes II
See more of this Group/Topical: Topical E: High Temperature Environmentally Sustainable Energy Processes (sessions joint with the Environmental Division)

# (391e) Modeling and Control of Chemical Looping Combustion Process

**Timothy P. McFarland**[1], Balaji Sukumar[1], B. Erik Ydstie[2] and Bruce Krogh[3], (1)Chemical Engineering, Carnegie Mellon University, Pittsburgh, PA, (2)Department of Chemical Engineering, Carnegie Mellon University, Pittsburgh, PA, (3)Electrical and Computer Engineering, Carnegie Mellon University, Pittsburgh, PA

Chemical looping combustion (CLC) is a promising clean technology for power generation. In this technology, metal oxides are looped between the oxidation and the reduction chambers thereby carrying oxygen from air to fuel. Thus, the combustion of fuel is accomplished without any direct contact between the fuel and the air. This prevents the formation of harmful gases and allows for the formation of pure CO2, making CLC a good candidate for CO2 sequestration. Much of the previous research in CLC has focused on the process chemistry of various metal oxide and fuel combinations. With the breadth of experimental data available, a mathematical model was previously developed to simulate the effects of process variables (metal oxides, fuel).1 The model uses a shrinking core mechanism and discretized population balances to simulate the solid-gas reactions. The mathematical model is useful for both design and control of the CLC process. Scaling up the CLC process from lab and bench scale experiments to a commercialized process requires a transition from simple control algorithms to a complex network of interacting controllers.

In the current study, we focus on proposing a suitable control strategy for the CLC process based on the proposed process model.1 The process relies on the use of two interconnected fluidized bed reactors with seals to maintain independent operation. However, the reactors must simultaneously operate with competing control objectives: 1) match power demand specifications; and 2) achieve complete oxidation of products. These objectives coupled with the fast natural system dynamics lead to a challenging control problem. Due to the interconnectivity of the two reactor systems and the steam cycle, decentralized control may lead to instabilities and poor system response. In such cases, the principles of inventory control for extensive variables: energy, entropy, and volume; and decomposition of the control system into a dynamic network can provide a stable design. In this work, based on the concept of inventories, we propose a robust control for the CLC process along with advanced control schemes for improved system response.

1. Balaji, S., Ilic, J., Ydstie, B. E., Krogh, B. H. Control based modeling and simulation of the chemical looping combustion process, Industrial Engineering & Chemistry Research, In press. DOI: 10.1021/ie901540m.

See more of this Session: Chemical Looping Processes II
See more of this Group/Topical: Topical E: High Temperature Environmentally Sustainable Energy

# (391f) Design Aspects and Insights From Experimental Data in Chemical Looping with Oxygen Uncoupling for Solid Fuel Combustion

Asad H. Sahir, Dept. of Chemical Engineering, Institute for Clean and Secure Energy, University of Utah, Salt Lake City, UT, Adel. F. Sarofim, Dept. of Chemical Engineering, University of Utah, Salt Lake City, UT and JoAnn Slama Lighty, Department of Chemical Engineering, Institute for Clean and Secure Energy, Salt Lake City, UT

Chemical Looping with Oxygen Uncoupling is a potential combustion method being researched for solid fuels, which would assist in the economic capture of carbon dioxide. An order of magnitude process engineering analysis has been developed based on existing literature and insights from in-house kinetic experiments conducted on copper oxide as an oxygen carrier. In this study, data from in-house kinetic experiments has been employed to establish criteria for appropriate temperature and residence time for the air and fuel reactor. The importance of an optimum recirculation of metal in the Chemical Looping with Oxygen Uncoupling system will also be discussed. The talk would also highlight on the utilization of global coal char oxidation kinetic studies from literature in conjunction with the time required for oxygen release by cupric oxide, for investigating the residence time requirements in the fuel reactor.

See more of this Session: Chemical Looping Processes II
See more of this Group/Topical: Topical E: High Temperature Environmentally Sustainable Energy Processes (sessions joint with the Environmental Division)

# (419a) Latest Developments at the Vattenfall Oxycoal Demonstration at the Schwarze Pumpe Plant

**Lars Stromberg**, Vattenfall AB, Stockholm, Sweden

Vattenfall AB, one of the largest utilities in the European Union (EU), has embarked on the first large-scale demonstration of oxy-coal combustion in the world at the Schwarze Pumpe plant in Germany. The pilot plant, in operation since the Fall of 2008, includes its own Air Separation Unit, a single, commercial-scale oxy-coal burner housed within a water-wall boiler, and a system for carbon dioxide purification and compression into liquid form prior to transport to a sequestration site. This presentation will discuss the latest developments at our pilot plant, as well as the development of our oxy-fuel commercial demonstration unite, and the associated engineering considerations.

See more of this Session: Oxycombustion of Coal - Need, Opportunities, and Challenges I
See more of this Group/Topical: Topical E: High Temperature Environmentally Sustainable Energy Processes (sessions joint with the Environmental Division)

# (419b) SO3 Emissions From a Tangentially-Fired Pilot Scale Boiler Operating Under Oxy-Combustion Conditions

Michael M. Clark, James R. Kenney, Armand A. Levasseur and Shin G. Kang, Boiler Combustion Systems, Alstom Power Inc., Windsor, CT

Oxy-combustion of pulverized coal (PC) is a promising near-term technology for CO2 capture and sequestration from utility boilers. With support from DOE, ALSTOM Power is executing an oxy-combustion test program utilizing the 15 MWth tangentially-fired Boiler Simulation Facility (BSF), in Windsor, CT, USA. The objectives of this program are to design and develop an innovative oxyfuel system for T-fired boiler units, evaluate the performance of the oxyfuel T-fired boiler system in pilot-scale tests, and address technical gaps for commercial design of oxyfuel systems for T-fired utility boilers. In evaluating the oxyfuel system, detailed measurements from the pilot-scale tests will aid in understanding furnace heat absorption, ash deposition, material corrosion, combustion, and pollutant formation and destruction in the oxyfuel T-fired process. This presentation will focus on the effort to understand the fate of SO3 under oxy-combustion conditions. Measurements were taken at various locations on the BSF during air-fired baseline condition and during two different oxy-combustion conditions to quantify the impact of flue gas recycle (FGR) on SO3 formation. Two different FGR conditions were evaluated for oxy-combustion operation: one with high SO2 in the FGR and one with very low SO2 in the FGR. Measurements of SO3 concentration were obtained by applying a controlled condensation method with flue gas sample extraction via an inertial separation probe. Insights into the formation and fate of SO3 under oxy-combustion conditions will be discussed.

See more of this Session: Oxycombustion of Coal - Need, Opportunities, and Challenges I
See more of this Group/Topical: Topical E: High Temperature Environmentally Sustainable Energy Processes (sessions joint with the Environmental Division)

# (419c) Fate of Sulfur Trioxide Under Oxy-Coal Combustion Conditions

**Jiyoung Ahn**, Liyong Wang, Ryan Okerlund and Eric G. Eddings, Chemical Engineering, University of Utah, Salt Lake City, UT

In coal combustion, most sulfur present in the fuel is oxidized to sulfur dioxide, with only a small portion being converted to sulfur trioxide. However, in oxy-fuel combustion, due to the increased oxygen concentration, it is possible that more of the fuel-bound sulfur can be converted to sulfur trioxide. Many studies to date have focused on the formation of NOx under oxy-coal conditions; however, the formation of sulfur trioxide under oxy-coal conditions is not well understood. Since sulfur trioxide can form sulfuric acid with water vapor in flue gas and subsequently lead to corrosion, it is important to understand the formation of sulfur trioxide during oxy-coal combustion.

The purpose of this study is to investigate the effect of various operating parameters in combustion on the formation of sulfur trioxide. Modified ASTM method D3226-73T (also called the Controlled Condensation Method) was used to measure sulfur trioxide in bench-scale and pilot-scale experiments. Experiments were performed at various temperatures to understand the relationship between the combustion temperature and the amount of sulfur trioxide formed. The effects of oxygen and carbon dioxide concentration on the amount of sulfur trioxide were investigated for both air and oxy-fired combustions by changing the concentration of oxygen in the oxidizer. Also, this study investigated the influence of limestone ($CaCO_3$) on the amount of sulfur trioxide emitted during both air-fuel and oxy-fuel combustion. Since the formation of sulfur trioxide is directly related to the concentration of sulfur dioxide, a decreased amount of sulfur dioxide due to limestone capture should, in principle, affect the concentration of sulfur trioxide. The experimental results presented will be explained in the context of the relevant sulfur chemistry, and compared with the outcomes predicted by thermochemical and kinetic calculations.

See more of this Session: Oxycombustion of Coal - Need, Opportunities, and Challenges I
See more of this Group/Topical: Topical E: High Temperature Environmentally Sustainable Energy Processes (sessions joint with the Environmental Division)

## Potential Impacts of Oxy-Combustion Retrofit on Boiler Tube Corrosion Rate

Andrew Fry (Principle Contact, fry@reaction-eng.com)
Bradley Adams (co-author, adams@reaction-eng.com)
Dave Swensen (co-author, swensen@reaction-eng.com)

Reaction Engineering International
77 West 200 South Suite 210
Salt Lake City, UT 84101
Ph: 801.364.6925 Fax: 801.364.6977

William Cox (co-author, cox@corr-mann.com)
Corrosion Management
21 Sedlescombe Park
Rugby, CV22 6HL
United Kingdom
Ph: 44 7966 528 051

### Introduction

Fireside tube corrosion is a topic of great concern for US utility boilers. Some furnaces have experienced local tube metal loss rates on the order of 2.5 mm/yr (100 mil/yr) and EPRI recently estimated that fireside corrosion costs the U.S. electric power industry up to $590 million per year and is responsible for approximately half of unscheduled outages in steam generation units [Syrett and Gorman, 2003]. During the last twenty years, the introduction of low NOx firing systems (low NOx burners and overfire air) and efforts to improve thermal efficiency through the use of higher pressure/temperature steam conditions have resulted in growing concern with fireside corrosion and have led to increasing pressure to improve materials technology. Recent focus has been on the introduction of oxy-fuel combustion, which has the advantage of reducing the volume of exhaust combustion gases, thereby simplifying the approach to reducing $CO_2$ emissions. Oxy-fuel combustion will also change the heat balance within the boiler, especially when applied within existing boiler units. Additionally, the prospect of combusting coal with oxygen and flue gas recycle has the potential to affect corrosion behavior in a number of different ways.

REI and Corrosion Management Ltd. have performed pilot-scale experimentation to elucidate the potential impacts of oxy-coal combustion retrofit of utility boilers on corrosion. The goals of these tests were two-fold:

1. To demonstrate differences, if any, in corrosion behavior of materials typically implemented in US utility boilers between air-coal and oxy-coal combustion conditions.
2. To develop a data set suitable for validation of mechanisms developed to predict the corrosion rate of heat transfer surfaces under air- and oxy-fired coal combustion conditions.

## Experimental

In order to obtain a real-time indication of corrosion, a measurement system based on electrochemical noise sensing was utilized [Bakker et al., 1992]. This sensing technique was performed on corrosion probes with sensor elements fabricated from materials representative of heat transfer surfaces in high temperature combustion applications. One waterwall material and three superheater tube materials were chosen that are typical of those used in US utility boilers. The specifications of the materials chosen are presented in Table 1.

**Table 1.** Specification of the materials used for corrosion probe sensor elements

| Material | SA 210 | T22 | P91 | 347H |
|---|---|---|---|---|
| HT Surface | Waterwall | Superheater | Superheater | Superheater |
| C, % | 0.07 | 0.11 | 0.10 | 0.048 |
| Si, % | 0.23 | 0.2 | 0.32 | 0.40 |
| Mn, % | 0.42 | 0.44 | 0.47 | 1.32 |
| Ni, % | - | - | 0.15 | 9.73 |
| Cr, % | - | 2.21 | 8.52 | 17.45 |
| Mo, % | - | 0.95 | 0.93 | - |
| S, % | 0.002 | 0.003 | 0.002 | 0.008 |
| P, % | 0.009 | 0.01 | 0.018 | 0.026 |
| Cu, % | - | - | 0.11 | - |
| Al, % | - | - | 0.01 | 0.005 |
| Nb/Cb, % | - | - | 0.07 | 0.63 |
| V, % | - | - | 0.22 | 0.078 |
| N, % | - | - | 0.044 | - |
| Ta, % | - | - | - | 0.02 |

The materials shown in Table 1 are arranged in order of increasing alloy content from left to right. The SA210 material is a low-carbon steel that is commonly used to fabricate waterwall tubes in utility boilers and was used for the waterwall sensor element in this investigation. The remaining three materials, T22, P91 and 347H, are commonly used in the construction of superheater tubing and each of these was used to create the sensor elements for superheater probes. The waterwall and superheater probes differ in their physical construction. The sensor elements of the waterwall probe are at the tip of the probe on the flat surface of the probe. When installed in a furnace, the probe is positioned so that the front face of the probe is flush with the plane of the wall of the furnace. The elements of the superheater probe are rings of tubing that are cut from actual boiler tube. The corrosion surface of these elements has the standard mill finish. The superheater sensor elements are situated at the end of a long, cooled stainless steel lance. When inserted into the furnace, the orientation of the sensor elements is that of a tube in cross flow, precisely simulating the condition of a typical superheater tube in a utility boiler. The four probes used for this investigation are pictured in Figure 1.

**Figure 1**. Waterwall and superheater sensor probes used for real-time corrosion monitoring

The platform for this corrosion investigation was the University of Utah's 1.5 MW furnace, the L1500. The corrosion sensor probe and furnace configuration for these tests is presented in Figure 2. The sensor elements of the waterwall probe were typically maintained at 720°F and the probe was installed in section 2 of the L1500 with gas temperatures about 2300°F. The sensor elements on the superheater probes typically were controlled at 910°F and were installed at a gas temperature of about 1800°F. To target this gas temperature, the superheater sensor probes were installed in section 10 of the L1500. The superheater probes were typically installed with the P91 probe on top, the 347H probe in the center and the T22 probe on bottom. For all of these tests, the firing rate was 3.5 MBtu/hr. For the majority of these tests, the burner was staged with a stoichiometric ratio of 0.9. The overall stoichiometric ratio was sufficient to produce 3.0% $O_2$ (dry) in the flue gas and the concentration is dependent on air or oxy-fired conditions. The over-fire air was introduced in section 6 of the furnace. This configuration and typical operating conditions result in reducing conditions at the waterwall probe and oxidizing conditions at the superheater probes.

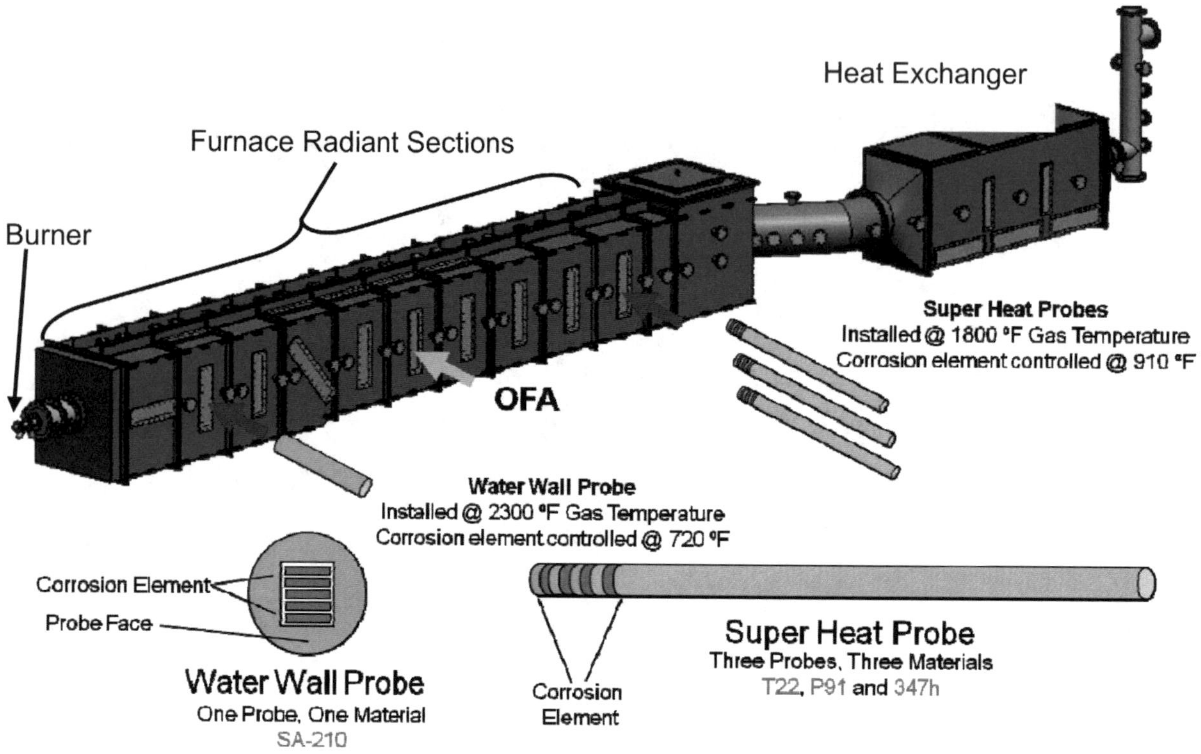


**Figure 2**. Pilot-scale furnace and corrosion sensor probe configuration

Three different coals were used for this experimental program. The Utah coal was from the Skyline mine and is a western bituminous, low-sulfur coal. The PRB coal was from the North Antelope mine and is also low sulfur. The Illinois coal was from the Shay #1 mine and has very high sulfur content. The analyses of the coals and of the mineral matter are presented in Table 2.

**Table 2**. Coal and mineral matter analyses

| | Skyline | PRB | Illinois |
|---|---|---|---|
| **Coal Analyses** | | | |
| C | 70.60 | 53.72 | 64.67 |
| H | 5.05 | 3.57 | 5.59 |
| N | 1.42 | 0.78 | 1.12 |
| S | 0.53 | 0.23 | 3.98 |
| O | 10.39 | 13.07 | 16.65 |
| Ash | 8.38 | 4.94 | 7.99 |
| Moisture | 3.18 | 23.69 | 9.65 |
| Volatile Matter | 38.60 | 33.36 | 36.78 |
| Fixed Carbon | 49.39 | 38.01 | 45.58 |
| HHV, Btu/lb | 12,606 | 9,078 | 11,598 |

| Mineral Matter Analyses | | | |
|---|---|---|---|
| **Al** | 14.52 | 14.78 | 17.66 |
| **Ca** | 6.11 | 22.19 | 1.87 |
| **Fe** | 5.09 | 5.20 | 14.57 |
| **Mg** | 1.39 | 5.17 | 0.98 |
| **Mn** | 0.02 | 0.01 | 0.02 |
| **P** | 0.59 | 1.07 | 0.11 |
| **K** | 0.57 | 0.35 | 2.26 |
| **Si** | 60.89 | 30.46 | 49.28 |
| **Na** | 1.41 | 1.94 | 1.51 |
| **S** | 2.33 | 8.83 | 2.22 |
| **Ti** | 0.88 | 1.30 | 0.85 |

* all values in mass % unless otherwise specified

## Results & Discussion

In order to determine the dependence of corrosion rate, several parameters were varied in these experiments. These parameters included: air- vs. oxy-fired conditions, coal sulfur concentration, corrosion sensor element temperatures and oxidizing vs. reducing conditions. However, over the six week period of in-furnace testing the conditions detailed in Figure 2, while firing with a burner stoichiometric ratio of 0.9, were maintained for a majority of the test program in order to confidently compare corrosion rate for air- and oxy-combustion.

The average corrosion rates calculated and reported here assume that the metal loss occurs uniformly over the surface of the corrosion sensor element. In reality, the loss of material is more likely to occur in discrete locations on the sensor element associated with deposit and flow characteristics. For the superheater probes this discrete area is probably limited to about a third of the overall surface on the leading and trailing sides of the tube in cross flow. Therefore peak corrosion rates for the superheat probes are likely to be approximately 3 times higher than the overall average rates reported here. However, as this is a pilot-scale facility which was operated only during an eight to ten hour period each day, and the corrosion probes were removed each night, the relative corrosion rate values were more relevant for comparison purposes than were the absolute rate values. The average corrosion rates measured for the baseline conditions described in Figure 2 are presented in Figure 3 on a large scale and in Figure 4 on a small scale.

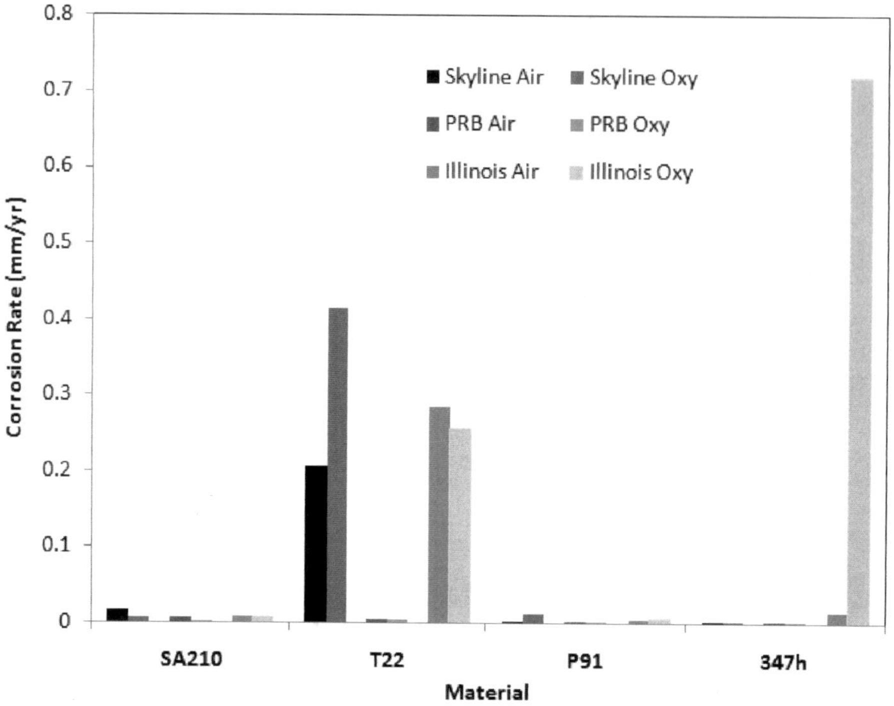


**Figure 3.** Summary of average corrosion rates for the baseline sensor element temperatures, probe locations and staging conditions (large scale)

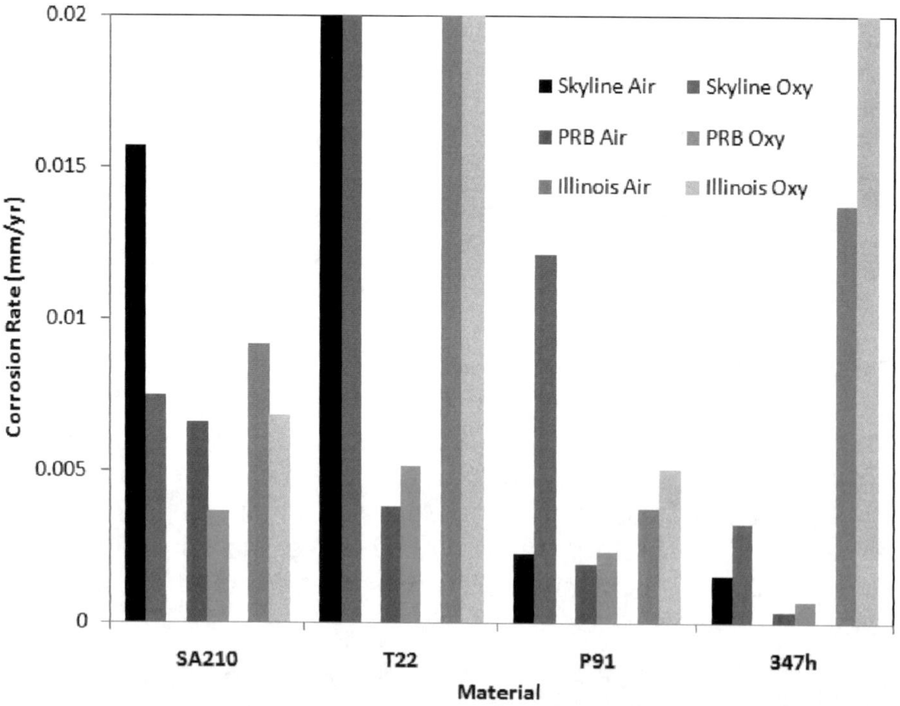


**Figure 4.** Summary of average corrosion rates for the baseline sensor element temperatures, probe locations and staging conditions (small scale)

Figure 3 and Figure 4 indicate very low corrosion rates for all of the materials, coals and air and oxy-firing, with the exception of five conditions. These conditions included the T22 material with Skyline and Illinois coals and for both air and oxy-fired conditions and the 347H material under Illinois/Oxy conditions. These five conditions and material combinations produced rates on the order of 0.21 to 0.72 mm/yr (8.3 to 28.3 mils/year), which are significant. While firing the Illinois coal, the flue gas $SO_2$ concentrations were approximately 3,200 ppmv (wet) for air-fired conditions and 17,600 ppmv (wet) for oxy-fired conditions. Later experiments suggested that at these high $SO_2$ concentration conditions, condensation of trisulphate species may be responsible for the high corrosion rate at the baseline probe setpoint temperatures. The corrosion rate of the 347H probe was significantly lower for the Illinois/Oxy conditions and an elevated probe setpoint temperature of 1150 °F.

More important than the absolute corrosion rates is the trend between the air and oxy-fired conditions. For the waterwall probe (SA210), the corrosion rate decreased when the combustion conditions were changed from air- to oxy-firing for all conditions tested. By contrast, for all of the superheater probes (T22, P91 and 347H) the corrosion rate increased when combustion conditions in the furnace were changed from air- to oxy-firing for all but one condition tested. To examine these trends in more detail it is useful to look at the plot of the increase in corrosion rate when converting from air- to oxy-firing for all materials and conditions. This plot is presented in Figure 2.9.

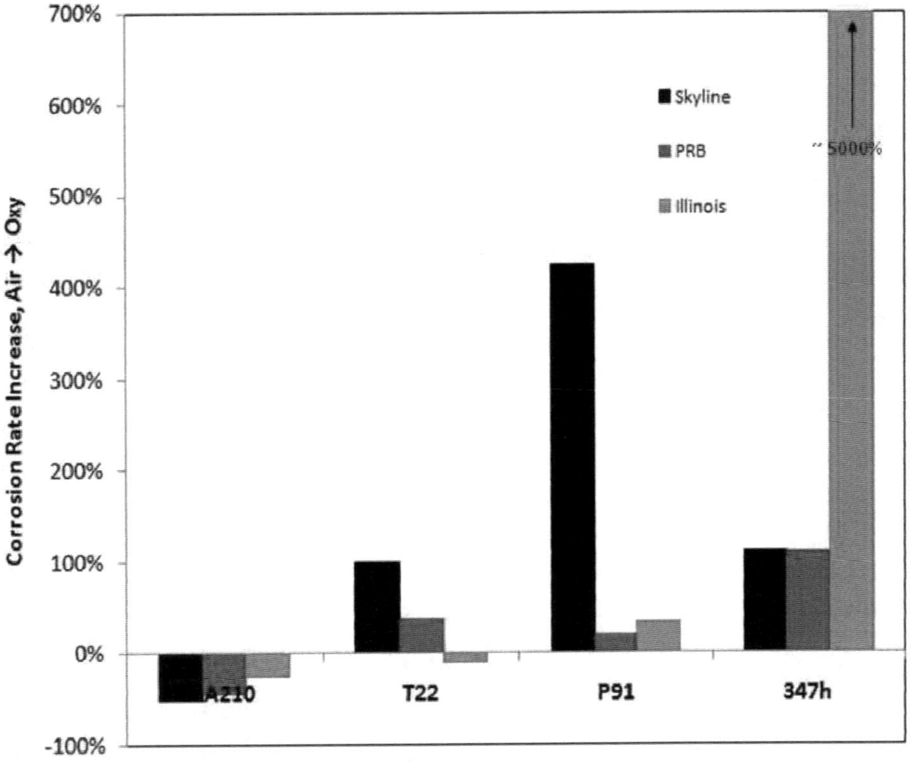


**Figure 5**. Measured increase in corrosion rate when converting from air- to oxy-firing

Figure 5 clearly indicates the difference in behavior between the waterwall and the superheater probes. The only material and coal that exhibited behavior that was contrary to the general trend was the T22 material with Illinois coal. The decrease in waterwall corrosion rate when converting from air to oxy-fired conditions, although involving minor corrosion rates, could be a significant circumstance. An expected decrease in corrosion rate to these heat transfer surfaces when retrofitting for oxy-combustion could be a facilitating result for the technology. The difference in behavior between the waterwall and superheater probes may be explained by the difference between oxidizing and reducing conditions. For these data, the waterwall probe was installed in a region with stoichiometry of 0.9, whereas the superheater probes experienced superstoichiometric conditions. The behavior of the superheater probes was unconcerning for most of the materials and coals tested. The increase in corrosion rates was small, compounding an already small corrosion rate. The results here that may be concerning and warrant further investigation are the P91 material with Skyline coal and again the 347H material with Illinois coal.

## Acknowledgments

This material is based upon work supported by the Department of Energy under Award Number DE-NT0005288. Project Manager is Timothy Fout. This report was prepared as an account of work sponsored by an agency of the United States Government. Neither the United States Government nor any agency thereof, nor any of their employees, makes any warranty, express or implied, or assumes any legal liability or responsibility for the accuracy, completeness, or usefulness of any information, apparatus, product, or process disclosed, or represents that its use would not infringe privately owned rights. Reference herein to any specific commercial product, process, or service by trade name, trademark, manufacturer, or otherwise does not necessarily constitute or imply its endorsement, recommendation, or favoring by the United States Government or any agency thereof. The views and opinions of authors expressed herein do not necessarily state or reflect those of the United States Government or any agency thereof.

## References

1. Bakker, W. T., Mok, W. Y., and Cox, W. M., "High-Temperature Fireside Corrosion Monitoring in the Superheater Section of a Pulverised-Coal-Fired Boiler", EPRI, Palo Alto, CA: 1992, TR-101799.
2. French, D. N., "Creep and Creep Failures", Bulletin of the National Board of Boiler and Pressure Vessel Inspectors, July, 1991.
3. Syrett, B. C., and Gorman, J. A., "Cost of Corrosion in the Electric Power Industry – An Update", *Materials Performance*, **42** (2), 32-38, 2003.
4. Viswanathan, R, and Bakker, W. T., "Materials for Boilers in Ultra Supercritical Power Plants", Proceedings of the International Joint Power Generation Conference, Miami Beach, Fl., July 23-26, 2000.

# (419e) Composition and Chemistry of Oxy-Coal Flames: a Discussion On Corrosion-Related Aspects

**Klas Andersson**[1], Daniel Fleig[2], Fredrik Normann[2] and Filip Johnsson[2], (1)Division of Energy Techology, Department of Energy and Environment, Chalmers University of Technology, Göteborg, Sweden, (2)Division of Energy Technology, Department of Energy and Environment, Chalmers University of Technology, Göteborg, Sweden

The present work addresses recent findings on the oxy-fuel combustion chemistry and the related composition of oxy-fuel flames. The two main topics are the sulphur chemistry occurring in coal combustion including $SO_2$, $SO_3$ and $H_2S$ formation and the formation and oxidation of CO in various coal flames. The implications on corrosion issues in oxy-fuel boiler design are discussed, as well as future research needs. The results presented include experimental work from the literature and results by the oxy-fuel group at Chalmers. The experiments at Chalmers are performed in a 100 kWth oxy-fuel test facility, which has been used extensively since its commissioning in 2003. In addition, modeling of the gas phase chemistry is discussed and related to the experimental work. All results presented for oxy-fuel combustion are related and compared against air-fired conditions.

The sulphur chemistry in oxy-fuel combustion differ from that under air-fired conditions; the main areas discussed in the paper are 1) conversion of fuel-S to $SO_2$, 2) $H_2S$ formation in the flame (at local fuel rich conditions) and 3) $SO_3$ formation. The $SO_2$ concentration increases drastically in oxy-fuel compared to air-fuel conditions. Both experiments and modeling show that the formation of $H_2S$ is promoted by the high $SO_2$ concentration in the oxy-fuel environment, given sub-stoichiometric conditions. The $SO_3$ concentration is also expected to increase, due to the elevated $SO_2$ concentration. Increased concentrations of $SO_3$ in combination with high moisture content in the flue gases (when recycling wet flue gases) will lead to a higher acid due point temperature with corrosion implications at low temperature parts of the boiler.

Both experimental and modeling work indicates that the CO formation is influenced by both homogeneous and heterogeneous reactions occurring in oxy-fuel combustion. Modeling of the homogenous chemistry in methane flames has shown that the main gas-phase reaction influencing the formation is

$CO_2 + H \leftrightarrow OH + CO$ (1)

Reaction 1 gives, besides a direct effect on CO formation, also an effect on the radical pool composition by promoting the formation of OH, which influences the $SO_3$ formation as is discussed further in the paper. High levels of CO are also found in experimental work on oxy-coal flames where gasification reactions are suggested to contribute to the CO formation, in addition to the homogenous reactions. The main reaction considered is the Boudouard reaction

$C(s) + CO_2 \leftrightarrow 2CO$ (2)

However, there are different opinions on the influence of gasification reactions on the CO formation in oxy-coal combustion, as will be discussed further in the paper. To summarize, high levels of CO and $H_2S$ are observed in the flame and near burner zone as well as higher $SO_2$ concentrations in the combustion zone in general. As mentioned, the $SO_3$ levels are expected to increase in oxy-fuel compared to air-fired conditions, but the formation of $SO_3$ needs to be studied in more detail under oxy-fuel conditions. For corrosion phenomena occurring at combustion temperatures, design principles otherwise employed during reburning and air/fuel staging in air-fired boilers may be adopted in oxy-fuel firing. For coals with high sulphur content and ashes with low sulphur retention capabilities, the temperature window of low temperature corrosion may be drastically influenced. Thus, oxy-fuel combustion is expected to cause a more corrosive environment than air-firing, due to the elevated concentration of combustion products.

See more of this Session: Oxycombustion of Coal - Need, Opportunities, and Challenges I
See more of this Group/Topical: Topical E: High Temperature Environmentally Sustainable Energy Processes (sessions joint with the Environmental Division)

2010 AIChE Annual Meeting

November 7-12, 2010

Salt Palace Convention Center

Salt Lake City, UT

# *Flexible Burner Concept for Oxyfuel Combustion*

Grathwohl S., Lemp, O., Müller M., Schnell U., Maier J., Scheffknecht G.[a],

Kluger F., Krohmer B., Moenckert P.[b]

[a] *Institute of Combustion and Power Plant Technology –(IFK), University of Stuttgart, Pfaffenwaldring 23,70569 Stuttgart, Germany*

[b] *Alstom Power Systems GmbH, Augsburger Str.712, 70329 Stuttgart, Germany*

---

Keywords: burner, oxygen injection, oxygen mixing , oxyfuel combustion

---

## ABSTRACT

Recently, $CO_2$ emission reduction or removal from coal fired power plants has become the highest priority in measures to combat global warming. Among researched technologies aiming at $CO_2$ capture in the power plant industry, the oxy-fuel process is nowadays of particular interest. In the oxy-fuel technology, the combustion of fuel is performed with pure oxygen and recycled flue gas instead of air; the firing system and in particular the burner is highly affected by the modified combustion process, which needs special attention. For testing purposes related to the oxy-fuel combustion process, a highly flexible firing system and burner concept is needed to cover the broad range of test conditions.

For the burner concept and the firing system one need to consider the following aspects:

- Ignition;
- Flame stability;
- Air operation mode;
- Oxyfuel operation modes
    - Direct oxygen injection mode
    - Pre-mixed oxygen mode
    - Individual oxygen enrichment mode;
- Switch from air to oxy-fuel mode and vice versa; and
- Several burner compartments and swirl implementation for burner tuning.

All these above mentioned aspects were considered by Alstom Power Systems GmbH in the design of the 30 MWth oxy-fuel burner of the Vattenfall's Schwarze Pumpe pilot steam generator.

In order to allow operation over a broad range oxyfuel conditions, this novel burner concept has been installed at the Institute of Combustion and Plant Technology (IFK) of University of Stuttgart in co-operation with Alstom Power Systems GmbH. Extensive tests have been carried out in a 500 $kW_{th}$ test facility and experimental data were compiled for mathematic modelling studies.

The downscale of the 30 MWth oxy-fuel burner to a size of 500kWth and the adaptation to boundary conditions of the IFK's 500kWth test rig as well as the testing of the burner have been performed within the framework of the European Commission funded project "OXYBURNER" (RFCR–CT2006–00009).

## INTRODUCTION

The IFK test facility is a vertical down-fired furnace with a total length of about 7 m and an inner diameter of 0.8 m, as illustrated in Figure 1. The test rig is optimized for investigation of pulverized fuel combustion processes and allows characterization of different kinds of fuels at staged and un-staged combustion conditions. The upper segments of the combustion chamber as well as the burner plate are equipped with refractory lining. A water jacket is integrated into the double-wall of the reactor. The test rig was retro-fitted to oxy-fuel operation by integration of a flue gas recycle path after the electrostatic precipitator. Pre-heating of either air or recycled flue gas is performed by using a heat exchanger installed downstream of the combustion chamber. For the oxy-fuel combustion experiments, pure oxygen from an external storage tank may either be premixed with the wet recycled flue gas or directly injected into the furnace. Bottled $CO_2$ and air is used as carrier gas for oxy-fuel and conventional operation modes, respectively. Accordingly, pure $CO_2$ or air serves as cooling/purge gas of the integrated flame detectors at the different operating conditions.

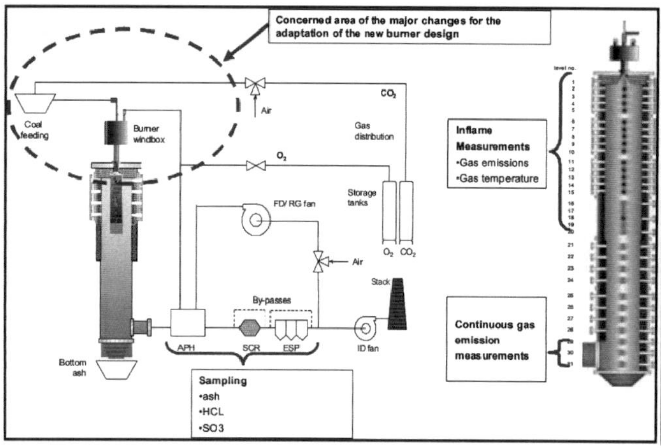


Figure.1: IFK 500kWth test facility

Figure.2: Implemented new burner

The downscaling task of the oxy-fuel burner and the adaptation to the 500kWth test rig's boundary conditions was supported by Alstom Power Systems GmbH. IFK has built, installed, and extensively tested the burner. Also, IFK produced experimental data for mathematic modeling studies.

The burner which was used has altogether five inlets, see Figure 2/3: three secondary inlets (core / V1 / V2), one for direct injection of oxygen (V3, designed as 8 lances), and the primary inlet (coal). The lances are arranged circularly around the primary air flow (Figure 3). The outer annular gap V1 provides a swirl generator. Hereby, the burner provides highest flexibility during operation, because the separate flows can be varied in volume flow as well as in $O_2$ concentration. Thus, making it possible for a direct injection of the entire not pre-mixed $O_2$ through the burner. The deflectors at the exhaust allow, even under varying volume flows and temperatures, a direct allocation and calculation of the impulses produced by the swirl. This is considered beneficial on the modelling process.

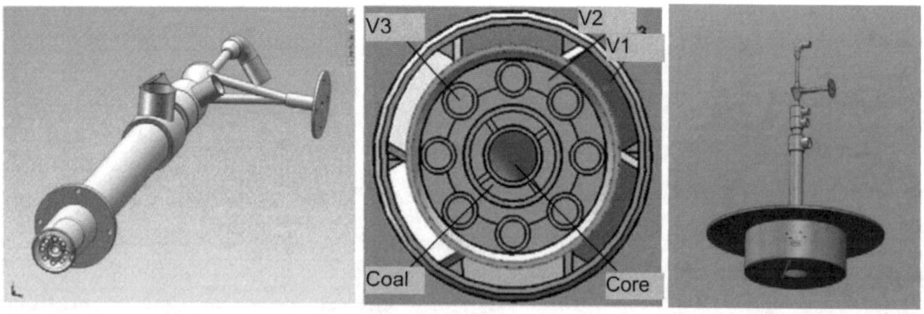


Figure 3: Design of the tested oxy-fuel burner of $0.5MW_{th}$ test rig

The experiments were conducted with pre-dried lignite. Different set-ups were tested as shown in Fig.4. The burner design offers the possibility to inject the necessary oxygen directly through the burner into the combustion zone. This allows varying the combustion conditions in a very wide range, up to the case where 100 % $O_2$ is directly injected.

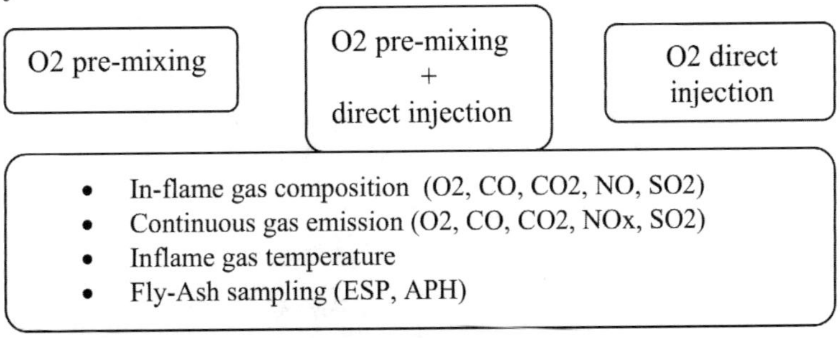


Figure 4: Possible conditions and measurements

During the experiments detailed in-flame measurements along the combustion chamber were done in order to characterize the flames resulting from different set-ups. The continuous monitoring of the gas emissions during the tests is a standard procedure. The coal quality (Lausitz TBK) which was used in the following combustion tests is listed in Table 1.

Table 1: Coal components

| | $H_2O$ | Volatiles | Ash | Fixed C | C | H | N | S | $O_{2\,(rest)}$ | $H_u$ |
|---|---|---|---|---|---|---|---|---|---|---|
| [%]an; [J/g]* | 10,7 | 48,0 | 4,9 | 36,4 | 56,2 | 4,7 | 0,5 | 0,5 | 22,5 | 21428* |
| [%]daf; | | | | | 66,6 | 5,6 | 0,6 | 0,6 | 26,7 | |

## TEST CASES

The test settings were called: Air, Oxy A, Oxy B, Oxy C (Table 2). As first operating status the air case (AIR) was initiated, from which the oxy-fuel cases were commenced. This allowed a stable and secure basic operation of the facility and the comparability from the oxy-fuel to the air case.

In the first setting under oxy-fuel conditions Oxy A the same adiabatic flame temperature as in the air case was aspired. Oxy B the amount of the recirculated flue gas was adjusted to keep the oxygen concentration in the whole combustion gas at about 21%.

In the third and last case (OXY C) for the first time, basing on the settings of Oxy B, the required oxygen amount was not premixed to the recirculated flue gas, but injected directly through the 8 oxygen lances of the burner into the combustion chamber. In both preceding cases, due to of safety reasons, a small part of the oxygen is channelled through the lances to avoid a return flow of the combustion products in those pipes and to provide cooling.

Table 2: Combustion conditions basic and advanced

| | Basic cases | | | Advanced case |
|---|---|---|---|---|
| | Air | Oxy A | Oxy B | Oxy C |
| Setting | | same adiabatic flame temperature as in air case | $O_2$, entirely 21% | Starting from Oxy 5 - $O_2$ directly into combustion chamber |
| power [$kW_{th}$] | 312 | 302 | 297 | 308 |
| Recirculation rate[1] [%] | ---- | 46 | 66 | 66 |
| Combustion gas [$Nm^3/h$] | 350 | 213,8 | 361 | 360 |
| $O_{2\,sec}$- burner [vol. % wet] | 20,9 | 40 | 25 | 5,5 |
| $O_{2\,total}$ [vol. % wet] | 20,9 | 33 | 20,5 | 20,7 |
| Swirl factor (calculated)) S | 0,26 | 0,08 | 0,24 | 0,17 |
| λ | ~1,19 | ~1,21 | ~1,25 | ~1,25 |

[1]pure $CO_2$ as buoyant gas / Flame detection

During the experiments detailed in-flame measurements along the combustion chamber were done in order to characterize the flames resulting from different set-ups. The continuous monitoring of the gas emissions during the tests is a standard procedure.

From profile measurements alongside the combustion chamber (Figure 5) the varying ignition behaviour of the individual settings can be determined.

The correlation between the calculated swirl factors and the measurement data is unambiguous, whereas Oxy 6 should be considered separately due to the $O_2$ direct injection.
The colored minima and maxima of the profiles (Figure 5 (a) and Figure 5 (b)) are scaled differently for a better illustration; therefore, a quantitative comparison is just limitedly possible.

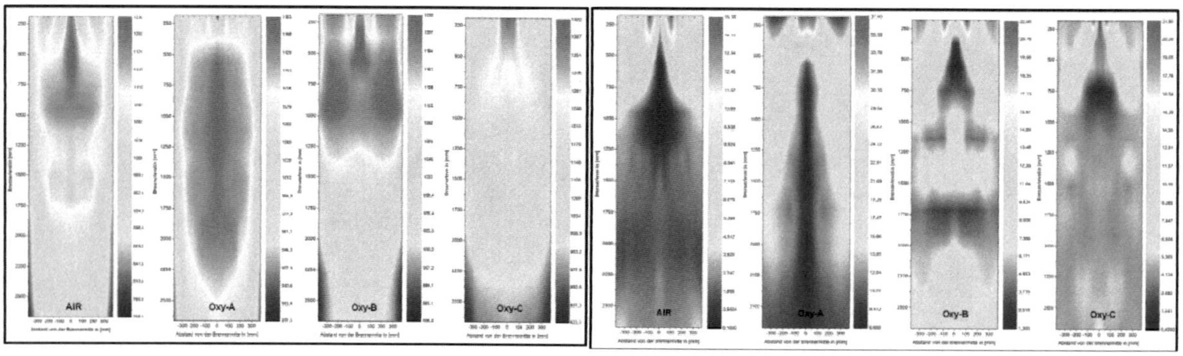


(a) (b)

Figure 5: Measured temperature profiles (a); Oxygen profile (b) (inflame measurement)

The emission measurements at the end of the combustion chamber (Table 3) indicate the high achieved $CO_2$ level in the flue gas as well as the already known increased pollutant concentrations due to the gas recirculation.
During the pollutant formation the CO concentration remained below the detection limit, for most of the oxy-fuel settings the values of the sulfur- and nitric oxides were significantly above those of the comparable air case. The $NO_X$ concentration reached values between 1000 and 1400 ppm (in comparison to 300 ppm in the air case) and the values of the $SO_2$ concentration were between 700 and 1100 ppm, in comparison to about 400 ppm during air combustion.
The measured combustion gas temperatures of the oxy-fuel cases varied from 1163 to 1422 °C compared to 1230 °C during the air combustion. The setting with the direct injection of the oxygen (Oxy C) shows the highest local temperatures, the one with constant adiabatic flame temperature (Oxy A) showed the lowest. According to that, the ignition behaviour in case of the direct injection is distinctly better. The following table shows the most important data of the emission gas measurement at the combustion chamber end.

Table 3: Emissions in volume concentration in flue gas

| | Air 2 | Oxy 4 | Oxy 5 | Oxy 6 |
|---|---|---|---|---|
| $O_2$ [vol. %, wet.] | 3,35 | 4,82 | 4,66 | 3,72 |
| $O_2$ [vol. %, dry.] | 3,78 | 6,75 | 6,63 | 5,57 |
| $CO_2$ [vol. %, dry.] | 15,9 | 89,6 | 89,7 | 90,3 |
| CO [vol. ppm, dry.] | < 10 | < 5 | < 5 | < 5 |
| $NO_X$ [vol. ppm, dry.] | 312,11 | 1099 | 1030 | 1318 |
| $SO_2$ [vol. ppm, dry.] | 413,98 | 1067 | 1089 | 768 |

The examination of the carbon presence in the fly-ash (ESP) (Table 4) shows excellent values for all the settings, which suggests an almost complete conversion of the combustibles. The oxy-fuel settings tend to result in even lower values. This correlates with the CO concentrations at the end of the combustion chamber.

Table 4: Carbon- und sulfur values in the examined E-filter ash samples

| | Air 2 | Oxy 4 | Oxy 5 | Oxy 6 |
|---|---|---|---|---|
| Carbon (C) [mass.-%] | 1,0 | 0,37 | 0,33 | n.b. < 0,3 |
| Sulfur (S) [mass.-%] | 3,1 | 6,2 | 6,2 | 6,8 |

The sulfur concentration of the fly-ash show in the oxy-fuel cases significant disparities in comparison to the air case. While the sulfur concentration of the ash in the air case is 3,1 w%, this value rises significantly in the oxy-fuel case up to 6,8 w%.

Those values are also coherent, compared with the measured $SO_2$ values at the end of the combustion chamber. The higher $SO_2$ partial pressure in relation with the used coal quality (ash composition) results in an increased sulphur integration.

## CFD SIMULATION

The simulations have been performed using the CFD code AIOLOS which is based on the Finite Volume method. Incompressible flow and stationary boundary conditions are assumed. The two-phase flow is treated as a simplified Eulerian quasi-one-phase flow neglecting any slip between gaseous phase and dispersed particle phase. The SIMPLE method provides the pressure-velocity coupling in combination with interpolation scheme from Date [ 1] for pressure correction. The standard k-" model is used to compute turbulence and the Eddy-Dissipation concept accounts for the influence of turbulence on chemical reactions. Modeling of radiative heat transfer is done by applying the discrete ordinates method. Details of the chemical reaction scheme and the radiation model is given in [2]. Moreover, the code has been optimized for vectorization and parallelization and allows domain decomposition. Hence, efficient use of supercomputers (i.e. high performance vector platforms) is enabled. Further information concerning AIOLOS is given elsewhere[3,4, 5].

The three-dimensional model used for the simulations consists of two separate domains:

- the burner domain with a cylindrical grid; and
- the furnace generated in Cartesian coordinates.

Details of the burner model are illustrated in Figure 5. The blue areas indicate the burner's inlet cross sections. The furnace domain was reduced to a length of about 4 m because the combustion process is essentially finished in this section, and no in-flame measurement data is available further downstream of the furnace. The entire mesh is discretized in approximately 2,200,000 computational cells.

Figure 5: Close-up of the computational grid.

The enhanced model which is based on the work of Leiser [6] has been validated against both an oxy-fuel and a conventional air-fired flame. The results of the simulations are summarized focusing on gas temperature profiles and the main gas species concentrations (O2, CO2, CO). The detailed results are given in [2]. In general, the predictive quality of the simulations appears to be quite satisfactory. Especially after a distance of approximately 0.5 m from the burner outlet, the simulation reflects the fundamental trends and fits very well with the experimental data. Nevertheless, discrepancies between measurements and simulation are identified in the burner near region.

## SUMMARY

With the new burner conception all of the defined oxy-fuel operation modes could be launched without any difficulties to an $O_2$ concentration of 21%. Furthermore, the tests showed that under oxy-fuel conditions a non pre-mixed operation where 100% of the oxygen is directly injected is possible. The emission and contaminant formation, at this volume flow and oxygen distribution, was already within the area of previous combustion experiments at IFK. The effect of the swirl number and momentum had a higher influence on the ignition and stability than the availability of oxygen in the reaction zone. The low CO values allow a reduction of the remaining oxygen concentration, whereby the $CO_2$ concentration in the exhaust gas increases further and the $NO_X$ concentration could be reduced simultaneously. The experimental data gathered at the test facility with the particular characteristics of the complex burner design can thus be reproduced by the numerical simulations with satisfactory accuracy. For the futher modeling process a database was created, on which optimizations can now take place. The experimental investigation supported by mathematical modeling studies provides valuable information on operational experience with different burner configurations.

## ACKNOWLEDGEMENT

The authors would like to thank the European Commission for greatfully funding the presented work within the projects OxyMod (RFCR-CT-2005-00006) and OxyBurner (RFCR-CT-2006-00009).

## REFERENCES

[1] A. W. Date, "Complete pressure correction algorithm for the solution of compressible navier-stokes equations on a nonstaggered grid," Numerical Heat Transfer, Part B: Fundamentals, vol. 29, no. 4, pp. 441–458, 1996.

[2] M. Müller, et al. "Advanced modeling of pulverized coal combustion under Oxy-fuel conditions," Clearwater Clean Coal Conference, 6 - 10 June 2010,Clearwater, USA.

[3] U. Schnell, "Numerical Modelling of Soild Fuel Combustion Processes using advanced CFD-based Simulation Tools," Progress in Computational Fluid Dynamics, vol. 1, no. 4, pp. 208–218, 2001.

[4] D. Förtsch, "A Kinetic Model of Pulverised Coal Combustion for Computational Fluid Dynamics," Ph.D. dissertation, Universität Stuttgart, Stuttgart, 2003.

[5] J. Ströhle, Spectral Modelling of Radiative Heat Transfer in Industrial Furnaces. Shaker Verlag, 2003.

[6] S. Leiser, "Numerical Simulation of Oxy-Fuel Combustion," Ph.D. dissertation, Universität Stuttgart, Stuttgart, to be published in 2010.

# (419g) Oxy-Coal Combustion with near Zero Flue Gas Recycle

Hisashi Kobayashi, Applications Research and Developement, Praxair, Inc., Danbury, NY and
Lawrence E. Bool, Applications Research and Development, Praxair, Inc., Tonawando, NY

Oxy-fuel combustion produces much higher adiabatic flame temperatures than air-fuel combustion. In order to convert an existing air fired boiler to oxy-coal firing flue gas recirculation (FGR) is required. In fact all of the proposed oxy-coal boiler conversion projects are based on FGR to control the flame temperature and to maintain proper balancing of heat transfer to different parts of the steam cycle. High adiabatic flame temperature and the resulting high heat flux are often erroneously cited as the reason for the FGR requirement. As discussed in this paper, the actual reason is the configuration of the existing air–fired boiler design, i.e., the radiant furnace section followed by a convective section, which has been optimized in the last century in order to maximize the steam cycle efficiency for air firing. For the construction of a new oxy-coal fired boiler, there are no fundamental heat transfer issues requiring FGR in the boiler design. In fact, there are over 1,000 industrial furnaces that have been successfully converted from air-fuel firing to full oxy-fuel firing without FGR. They are typically high temperature process furnaces such as steel reheating, glass melting, aluminum melting, copper melting and hazardous waste incineration furnaces. Precise control of furnace temperature and heat flux profiles in these process furnaces is often more critical than for utility boilers. Both high flame temperature oxy-fuel burners and advanced oxy-fuel burners utilizing internal flue gas recirculation were successfully applied. The heat flux profile was often improved after the oxy-fuel conversion by selecting a proper burner design and optimizing the burner placement and the firing rate distribution without any changes in the furnace geometry. These commercial examples have clearly demonstrated that oxy-fuel combustion can reproduce the desired temperature and heat flux profiles in any air-fuel fired high temperature furnaces. An oxy-coal firing system that does not require flue gas is often called 'zero recycle oxy-coal combustion'. However, safety and operational concerns may require some recirculated flue gas to transport the pulverized coal into the boiler. Therefore a more accurate expression is 'near zero recycle oxy-coal combustion'. In order to be viable, the boiler and burner designs must mitigate the potential drawbacks associated with high flame temperatures and high pollutant concentrations. For example the design must manage the heat flux profile to avoid extreme material temperatures and slagging and corrosion problems. The design would also need to be able to handle high concentrations of pollutants such as SOx and particulates that could potentially lead to high corrosion and slagging/fouling rates. Finally, the significant equipment and design modifications required to mitigate these potential problems make it likely that near zero recycle boilers will be new build rather than retrofits. In this paper previous studies on oxy-fuel combustion for boilers without FGR will be reviewed first and key boiler design issues with near zero FGR will be discussed. Several new oxy-fuel boiler design concepts without FGR are discussed.

See more of this Session: Oxycombustion of Coal - Need, Opportunities, and Challenges I

# Effects of Oxygen Concentration and Coal Composition on Aerosol Chemistry in Oxy Firing

**William J. Morris**
**Dunxi Yu**
**Jost O. L. Wendt**
Department of Chemical Engineering
University of Utah, Salt Lake City, UT 84112

**2010 AIChE Annual Meeting**

**Salt Lake City, Utah**

**November 7-12, 2010**

# Outline

- Objectives
- Coals examined
- Furnace, sampling, and analysis
- Results
- Conclusions

# Objectives

- Provide a comparison of three different coal aerosols under two different oxy fired scenarios for predicting effects of coal composition on oxy firing.
- Examine aerosol chemistry under two different oxy fired scenarios: 27% $O_2$ (~match heat flux of air) and 32% $O_2$ (~match adiabatic flame temperature of air).
- Use aerosol chemistry to provide information for those who wish to make predictions of fouling/slagging within the furnace.
- Determine whether there are any significant differences in aerosol chemistry at varying $O_2$ concentrations at conventional temperatures.

# Coal Chemistry

| Coal Analysis (on an as-received basis) | | | | | | | | | | | |
|---|---|---|---|---|---|---|---|---|---|---|---|
| **Sample** | **LOD** | **Ash** | **C** | **H** | **N** | **S** | **O (diff)** | **Volatile Matter** | **Fixed Carbon** | **HHV BTU/lb** | |
| | % | % | % | % | % | % | % | % | % | | |
| **Utah** | 3.18 | 8.83 | 70.60 | 5.41 | 1.42 | 0.53 | 13.21 | 38.60 | 49.39 | 12606 | |
| **PRB** | 23.69 | 4.94 | 53.72 | 6.22 | 0.78 | 0.23 | 34.11 | 33.36 | 38.01 | 9078 | |
| **Illinois** | 9.65 | 7.99 | 64.67 | 5.59 | 1.12 | 3.98 | 16.65 | 36.78 | 45.58 | 11598 | |
| **Ash Analysis** | | | | | | | | | | | |
| | **Al** as $Al_2O_3$ | **Ca** as CaO | **Fe** as $Fe_2O_3$ | **Mg** as MgO | **Mn** as MnO | **P** as $P_2O5$ | **K** as $K_2O$ | **Si** as $SiO_2$ | **Na** as $Na_2O$ | **S** as $SO_3$ | **Ti** As $TiO_2$ |
| | % | % | % | % | % | % | % | % | % | % | % |
| **Utah** | 14.52 | 6.11 | 5.09 | 1.39 | 0.02 | 0.59 | 0.57 | 60.89 | 1.41 | 2.33 | 0.88 |
| **PRB** | 14.78 | 22.19 | 5.2 | 5.17 | 0.01 | 1.07 | 0.35 | 30.46 | 1.94 | 8.83 | 1.3 |
| **Illinois** | 17.66 | 1.87 | 14.57 | 0.98 | 0.02 | 0.11 | 2.26 | 49.28 | 1.51 | 2.22 | 0.85 |

# Coal Firing Rates and Combustion Conditions

| Coal | Coal feed rate (kg/h) | Coal firing rate (kW) |
|---|---|---|
| **Utah Skyline** | 4.50 | 36.64 |
| **PRB** | 6.26 | |
| **Illinois Bituminous** | 4.89 | |

| | Combustion Mode | Oxidant Composition | Practical Significance |
|---|---|---|---|
| OXY27 | Oxyfuel Combustion | 27%$O_2$/73%$CO_2$ | Match AIR gas radiant heat flux |
| OXY32 | Oxyfuel Combustion | 32%$O_2$/68%$CO_2$ | Match AIR adiabatic flame temperature |

# Sampling Systems

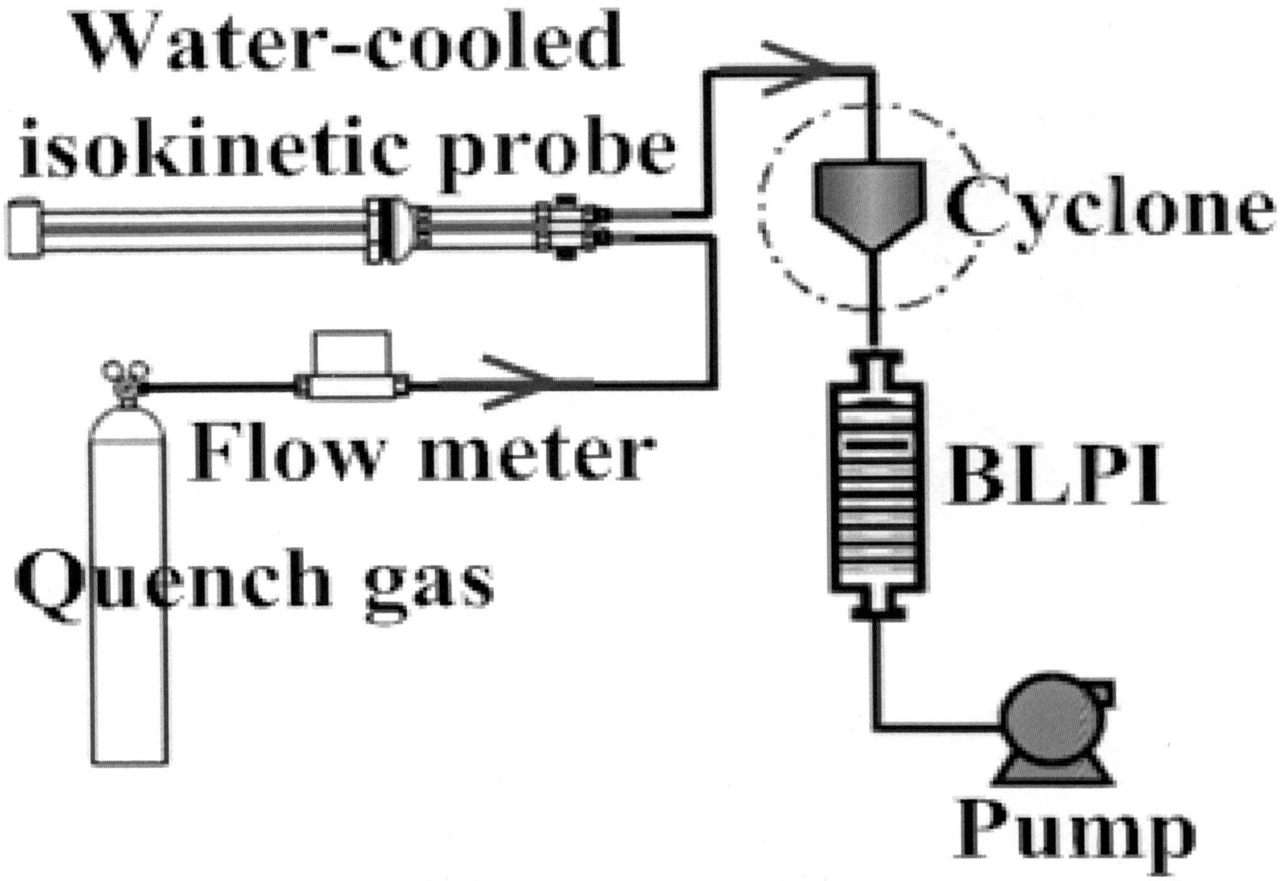

# Laboratory Combustor

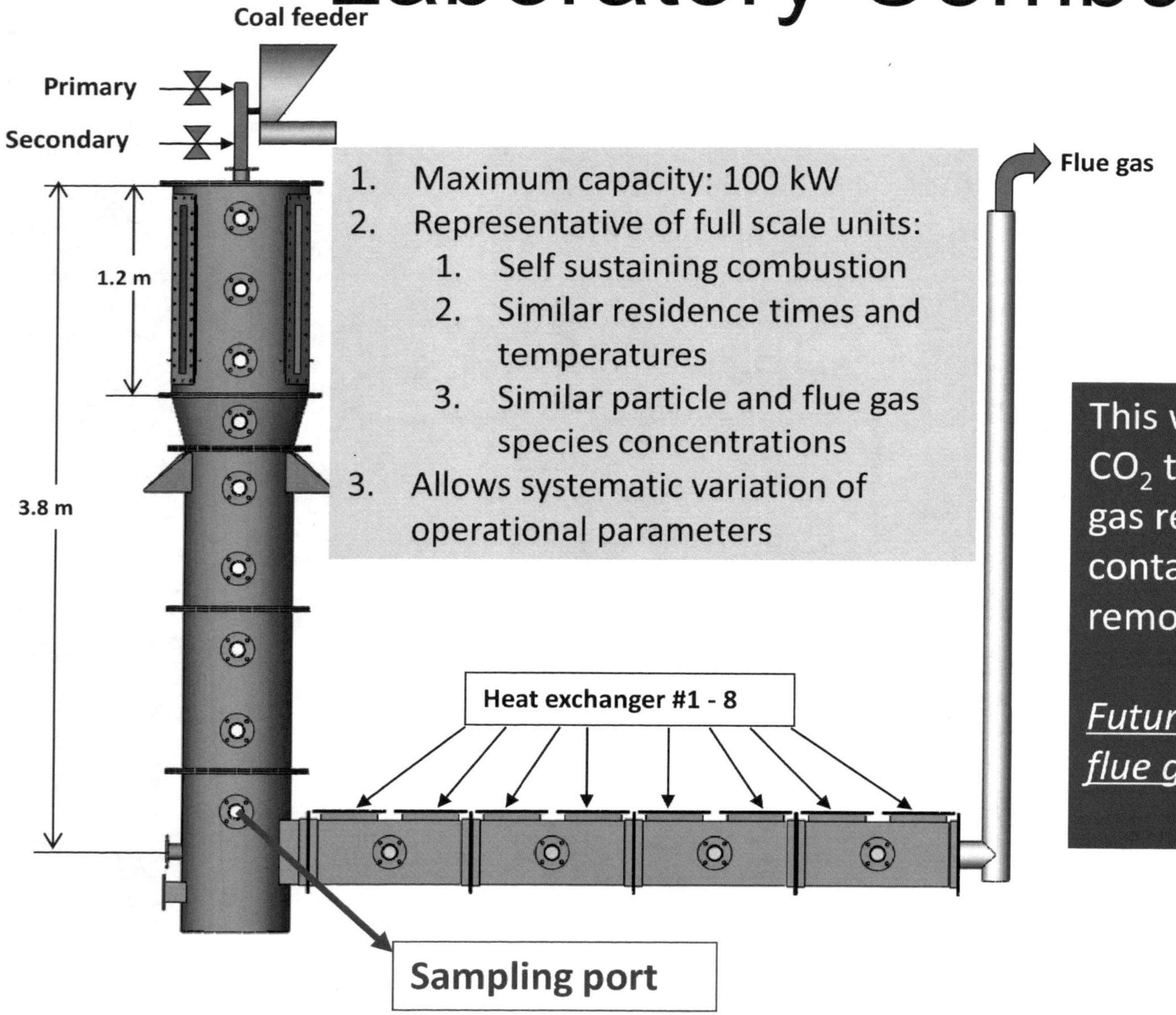


1. Maximum capacity: 100 kW
2. Representative of full scale units:
   1. Self sustaining combustion
   2. Similar residence times and temperatures
   3. Similar particle and flue gas species concentrations
3. Allows systematic variation of operational parameters

This work: Uses once-through $CO_2$ to simulate cleaned flue gas recycle with all contaminants and water removed.

*<u>Future work: Will use recycled flue gas.</u>*

# Analysis

- Samples were collected with a Berner low pressure impactor.
- Size segregated samples were then analyzed using ICP-MS.
- Due to the difference in flue gas volume (m3) of OXY27 and OXY32 conditions, results were also normalized to ug per g of coal burned.

# Results

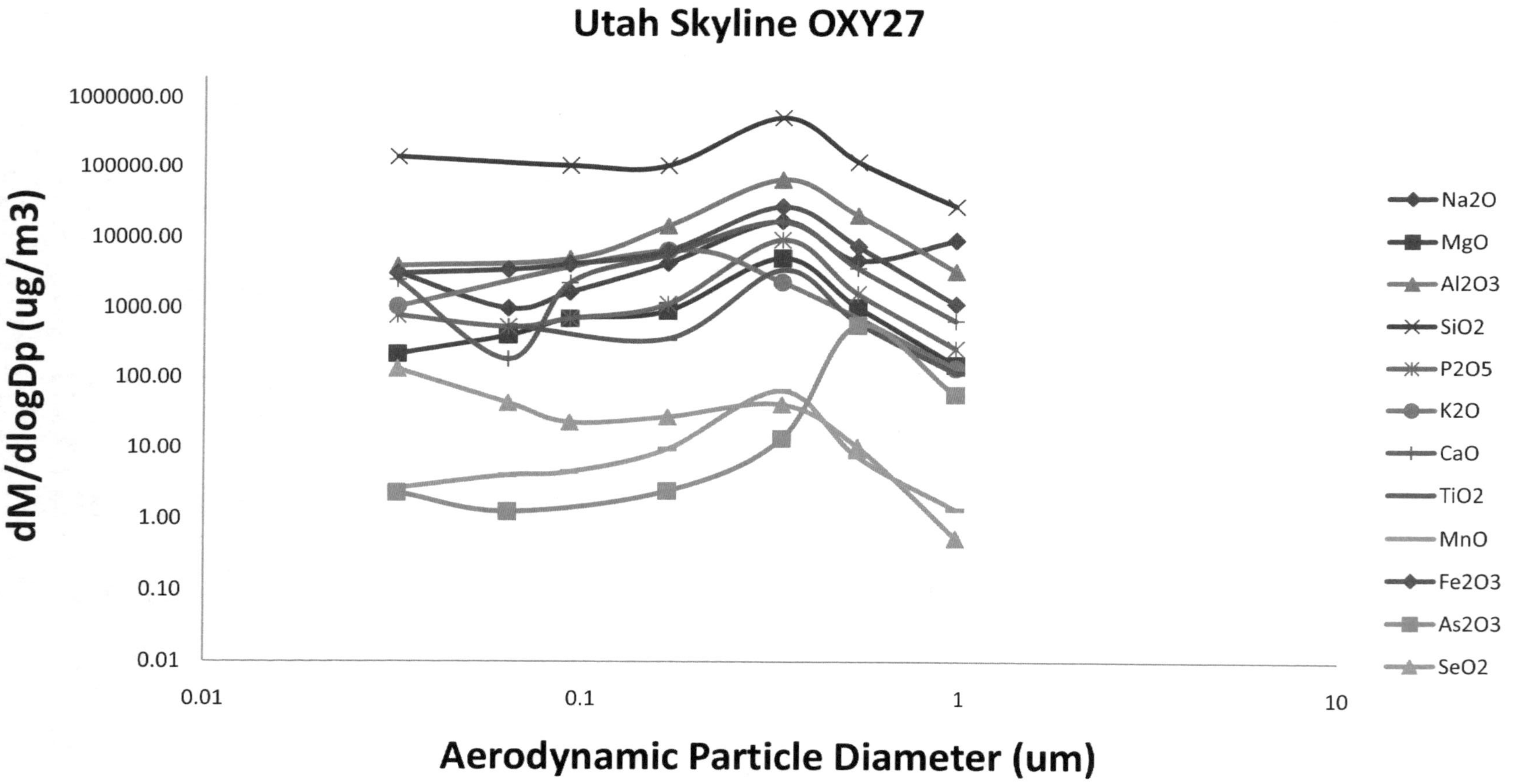
Utah Skyline OXY27
dM/dlogDp (ug/m3)
1000000.00
100000.00
10000.00
1000.00
100.00
10.00
1.00
0.10
0.01
0.01
0.1
1
10
Aerodynamic Particle Diameter (um)
Na2O
MgO
Al2O3
SiO2
P2O5
K2O
CaO
TiO2
MnO
Fe2O3
As2O3
SeO2

# Results

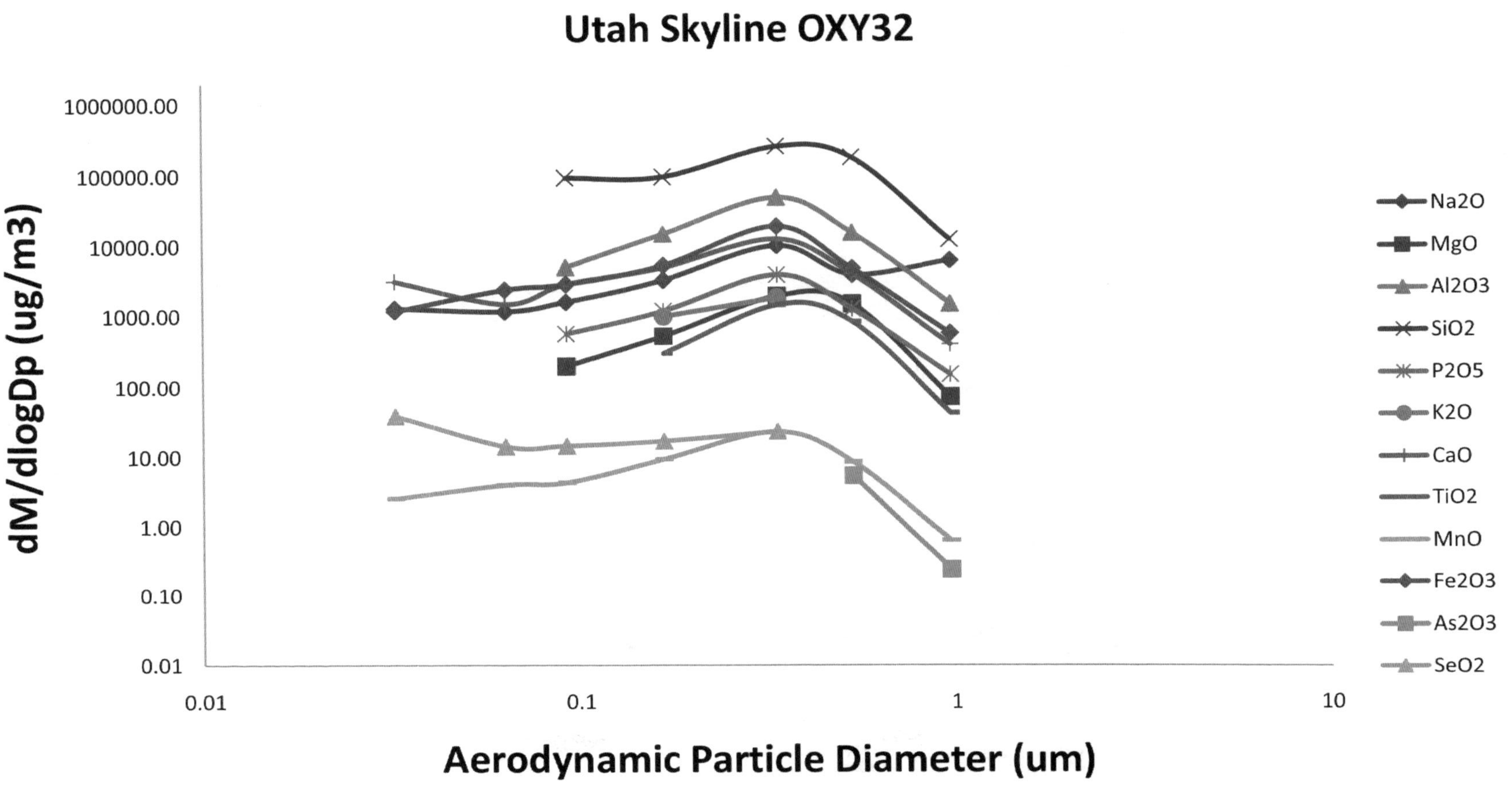
Utah Skyline OXY32
dM/dlogDp (ug/m3)
1000000.00
100000.00
10000.00
1000.00
100.00
10.00
1.00
0.10
0.01
0.01
0.1
1
10
Aerodynamic Particle Diameter (um)
Na2O
MgO
Al2O3
SiO2
P2O5
K2O
CaO
TiO2
MnO
Fe2O3
As2O3
SeO2

# Results

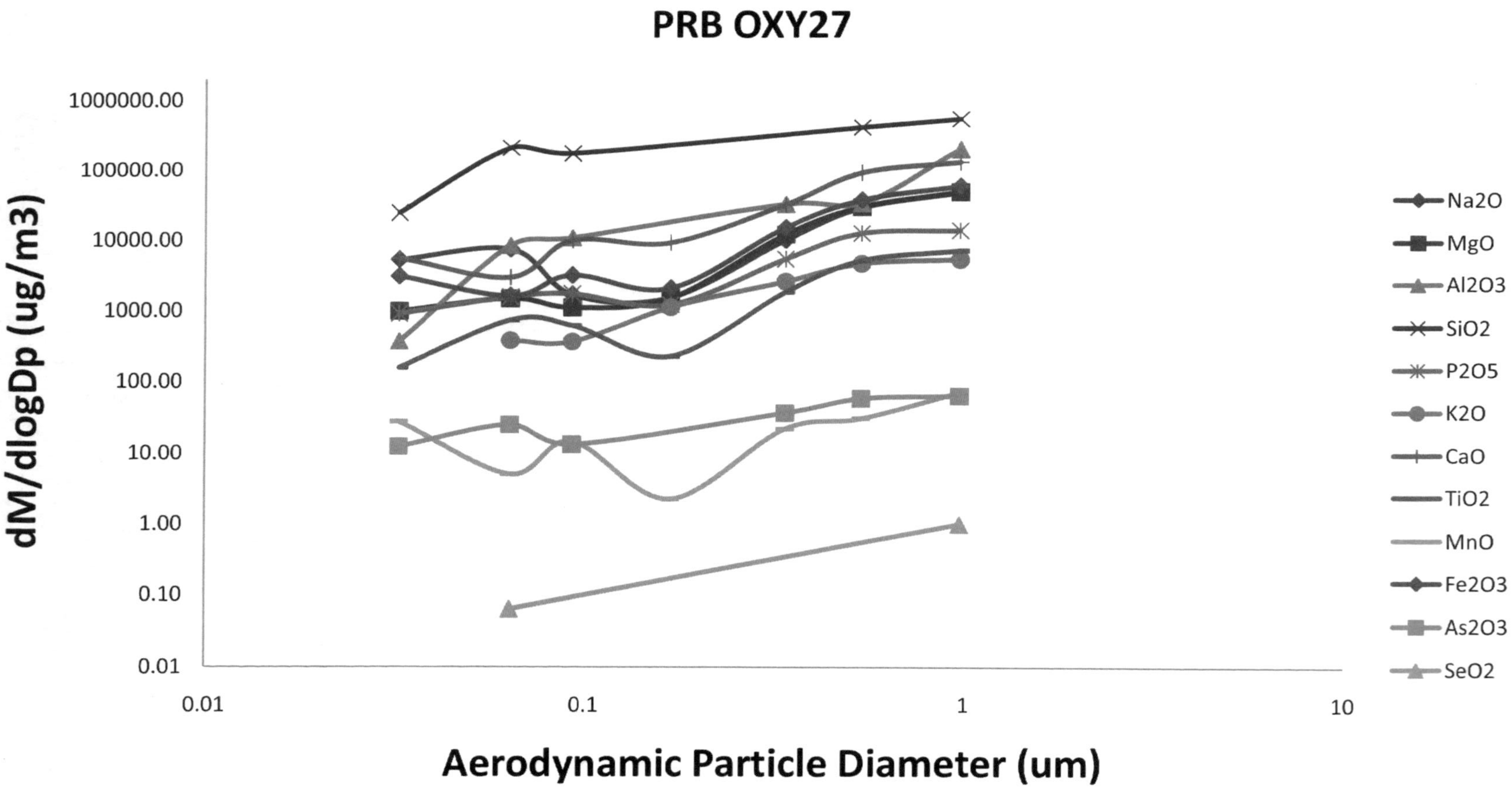
PRB OXY27
dM/dlogDp (ug/m3)
1000000.00
100000.00
10000.00
1000.00
100.00
10.00
1.00
0.10
0.01
0.01
0.1
1
10
Aerodynamic Particle Diameter (um)
Na2O
MgO
Al2O3
SiO2
P2O5
K2O
CaO
TiO2
MnO
Fe2O3
As2O3
SeO2

# Results

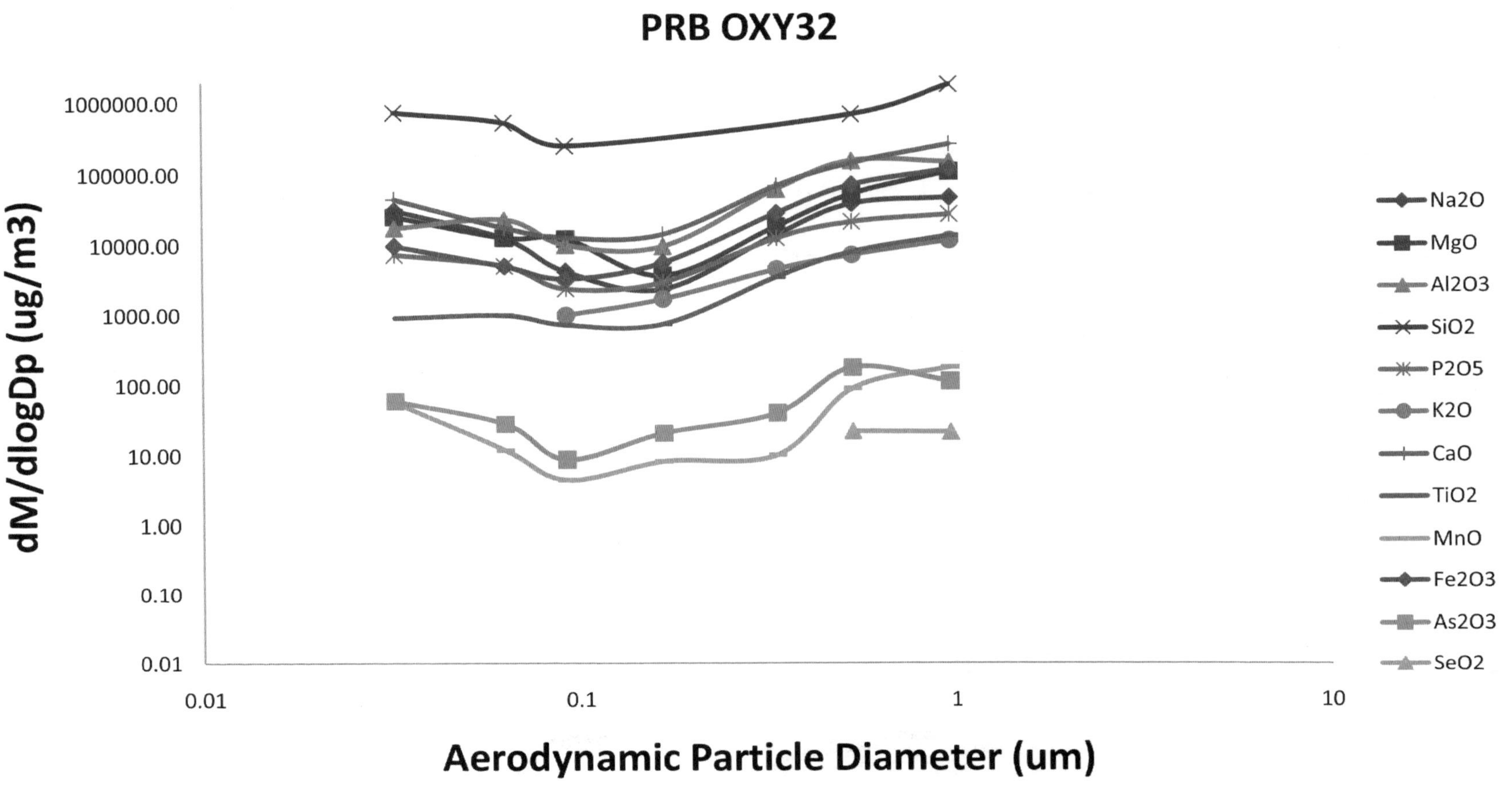

PRB OXY32
dM/dlogDp (ug/m3)
1000000.00
100000.00
10000.00
1000.00
100.00
10.00
1.00
0.10
0.01
0.01
0.1
1
10
Aerodynamic Particle Diameter (um)
Na2O
MgO
Al2O3
SiO2
P2O5
K2O
CaO
TiO2
MnO
Fe2O3
As2O3
SeO2

# Results

Illinois OXY27

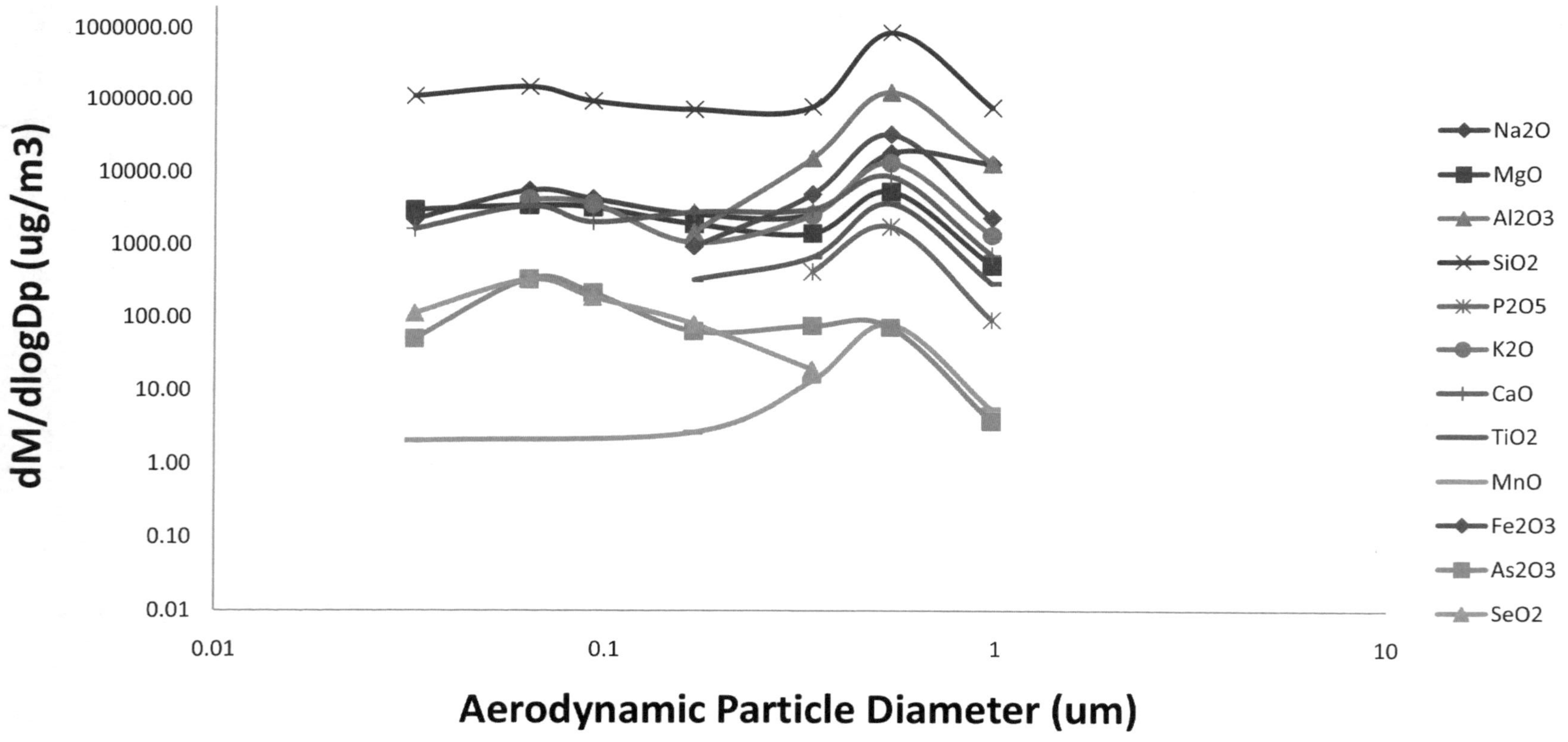

# Results

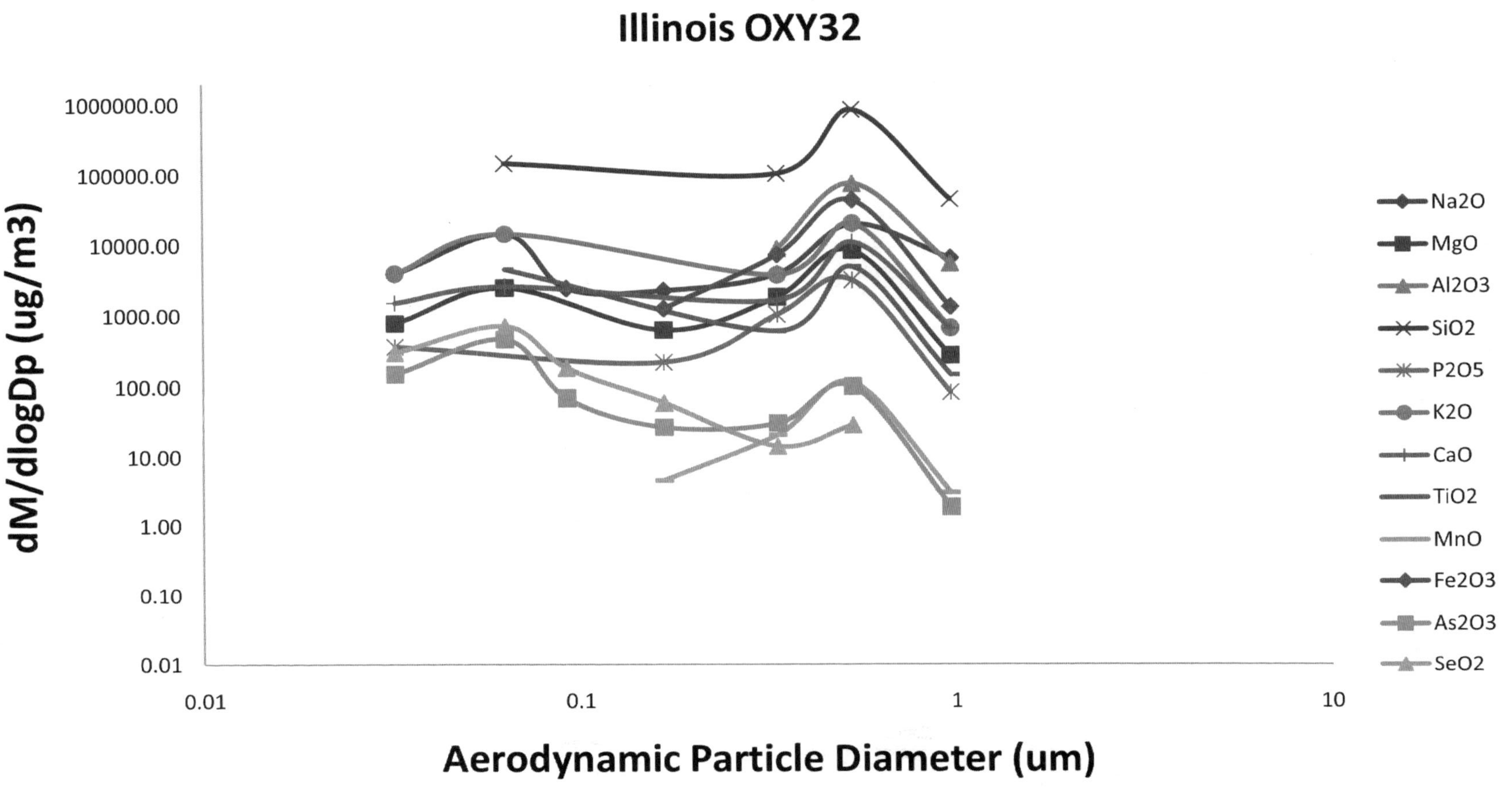
Illinois OXY32
1000000.00
100000.00
10000.00
1000.00
100.00
10.00
1.00
0.10
0.01
dM/dlogDp (ug/m3)
0.01
0.1
1
10
Aerodynamic Particle Diameter (um)
Na2O
MgO
Al2O3
SiO2
P2O5
K2O
CaO
TiO2
MnO
Fe2O3
As2O3
SeO2

# Results

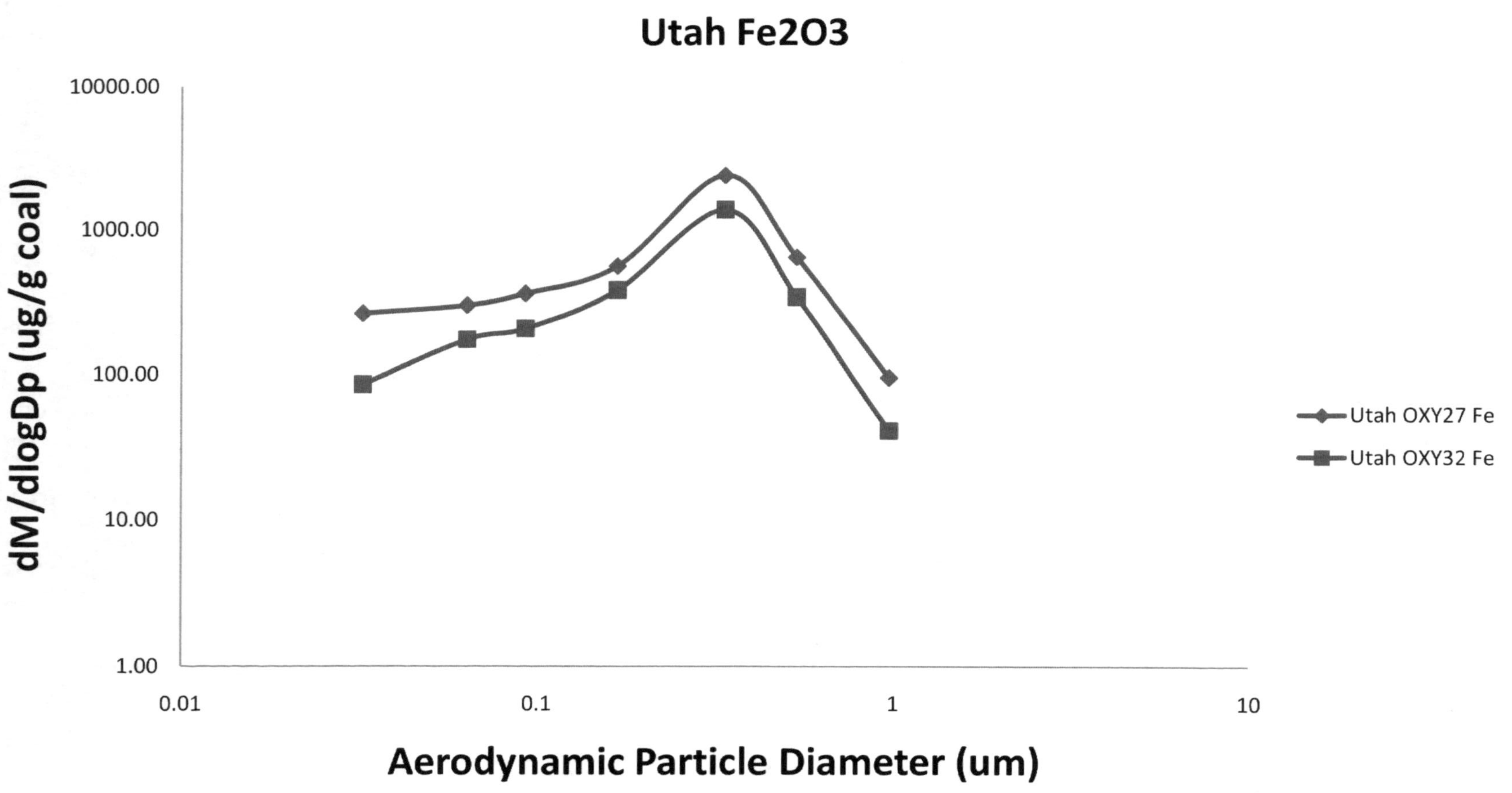
Utah Fe2O3
10000.00
1000.00
100.00
10.00
1.00
dM/dlogDp (ug/g coal)
0.01
0.1
1
10
Aerodynamic Particle Diameter (um)
Utah OXY27 Fe
Utah OXY32 Fe

# Results

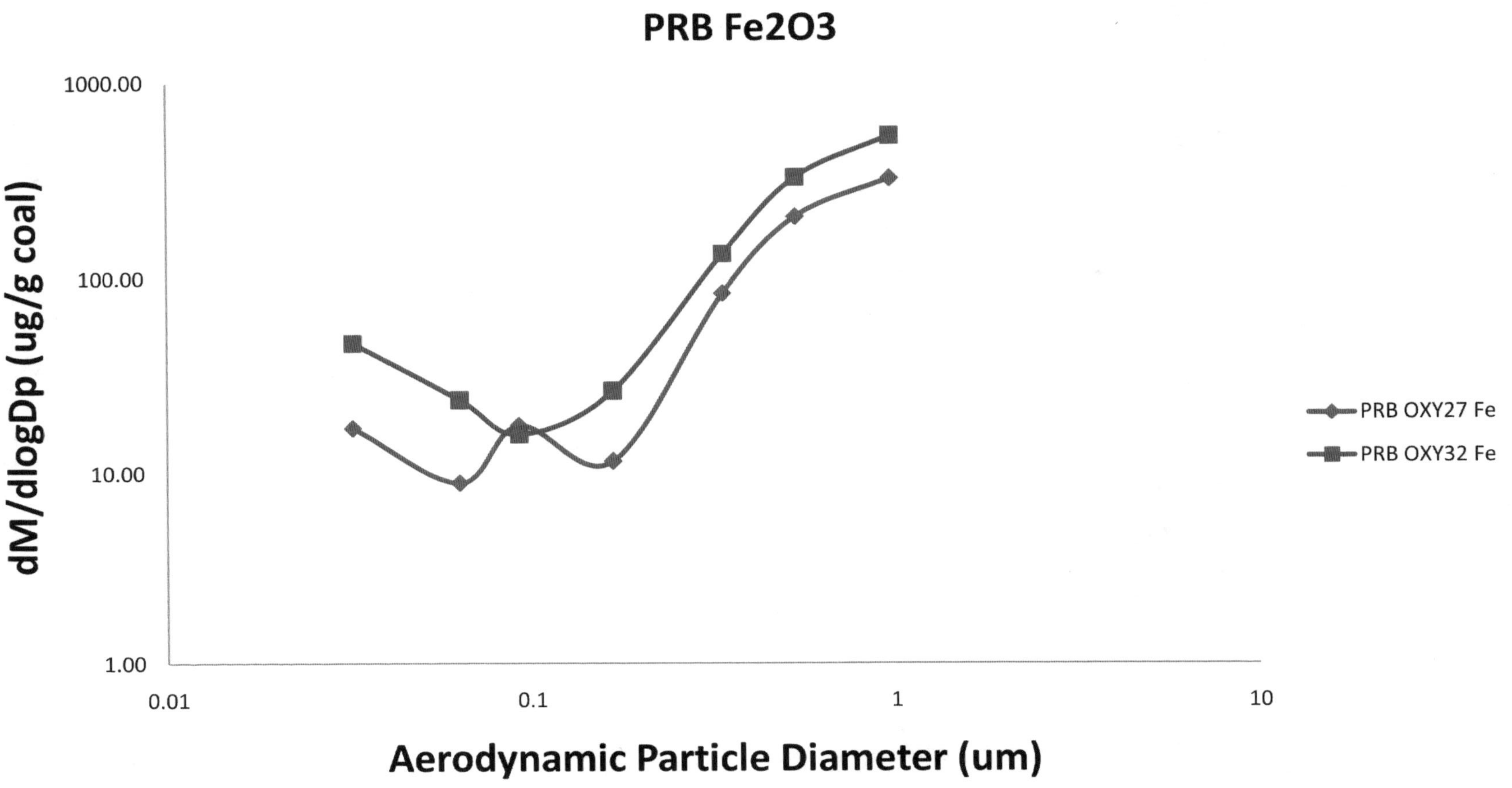

PRB Fe2O3
1000.00
100.00
10.00
1.00
dM/dlogDp (ug/g coal)
0.01
0.1
1
10
Aerodynamic Particle Diameter (um)
PRB OXY27 Fe
PRB OXY32 Fe

# Results

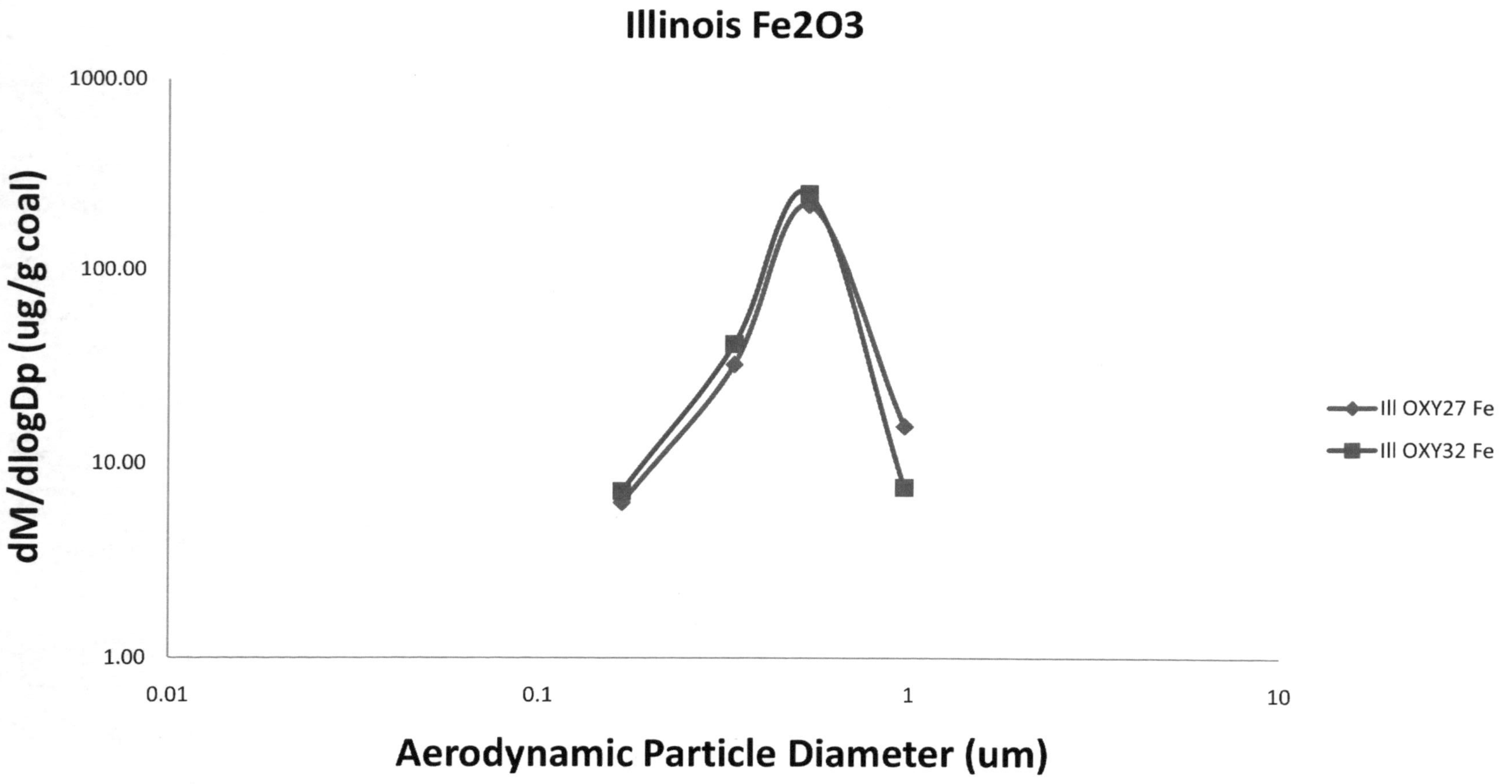
Illinois Fe2O3
1000.00
100.00
10.00
1.00
dM/dlogDp (ug/g coal)
0.01
0.1
1
10
Aerodynamic Particle Diameter (um)
Ill OXY27 Fe
Ill OXY32 Fe

# Results

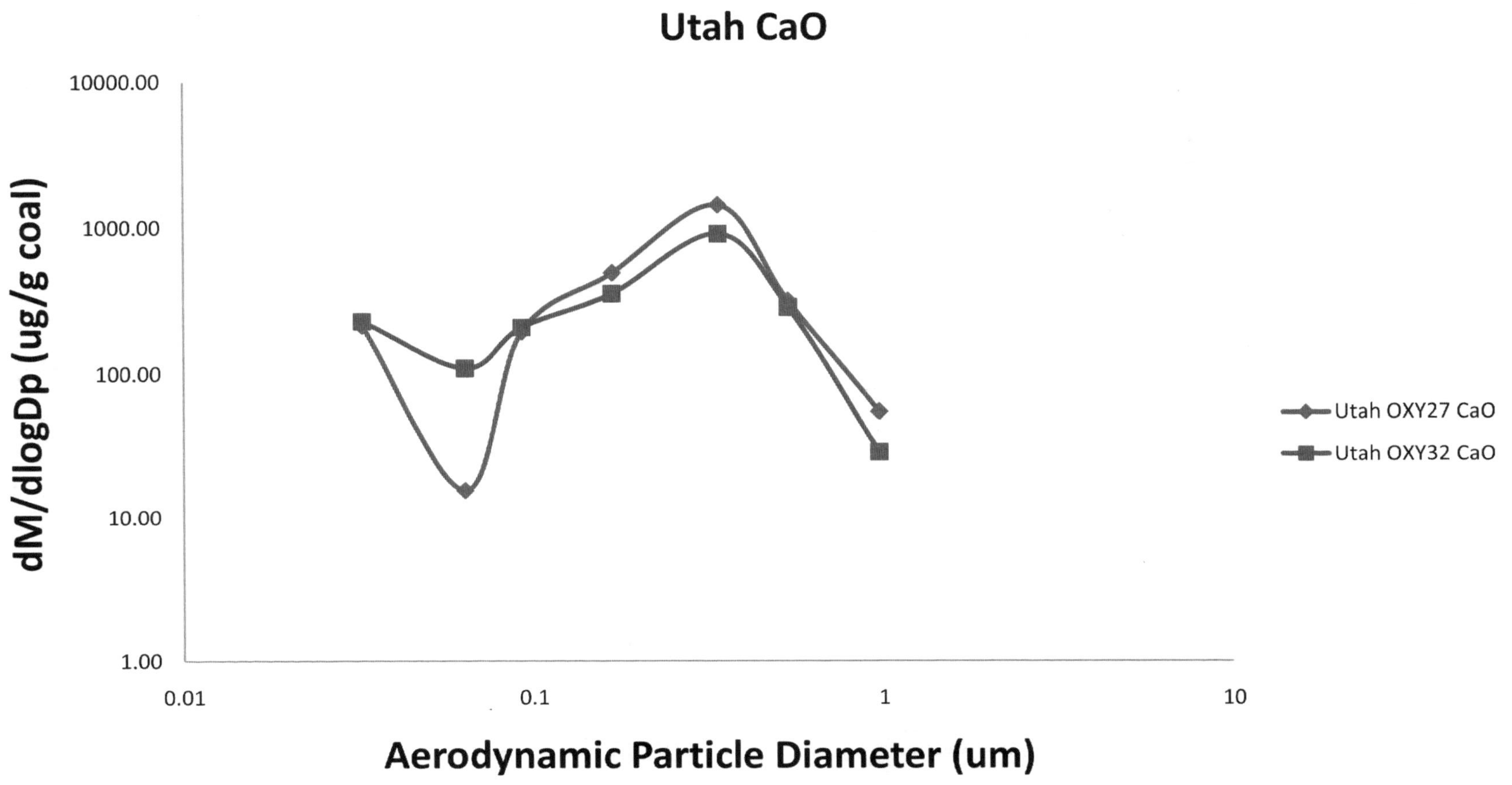

Utah CaO
10000.00
1000.00
100.00
10.00
1.00
dM/dlogDp (ug/g coal)
0.01
0.1
1
10
Aerodynamic Particle Diameter (um)
Utah OXY27 CaO
Utah OXY32 CaO

# Results

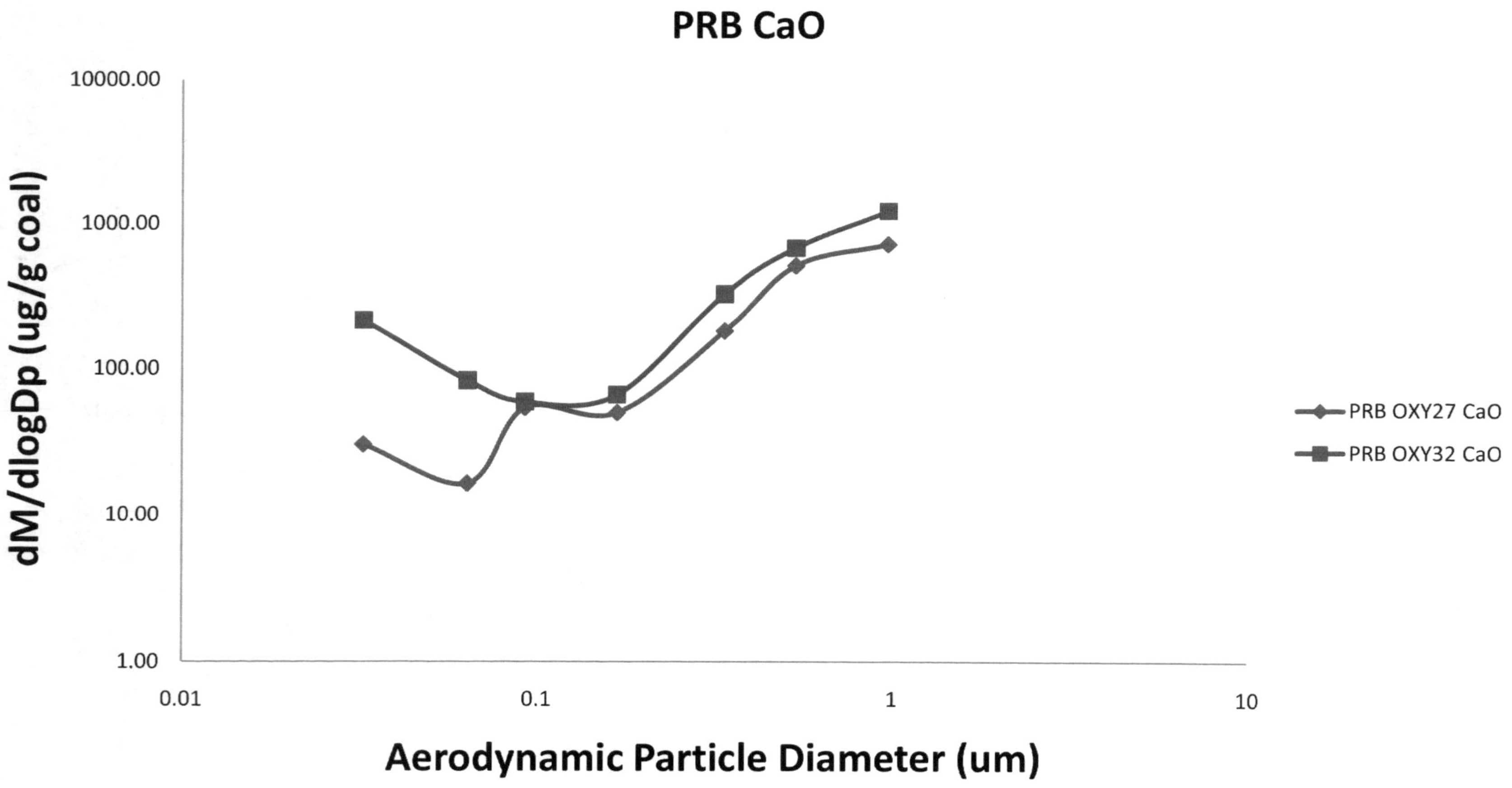
PRB CaO
10000.00
1000.00
100.00
10.00
1.00
0.01
0.1
1
10
dM/dlogDp (ug/g coal)
Aerodynamic Particle Diameter (um)
PRB OXY27 CaO
PRB OXY32 CaO

# Results

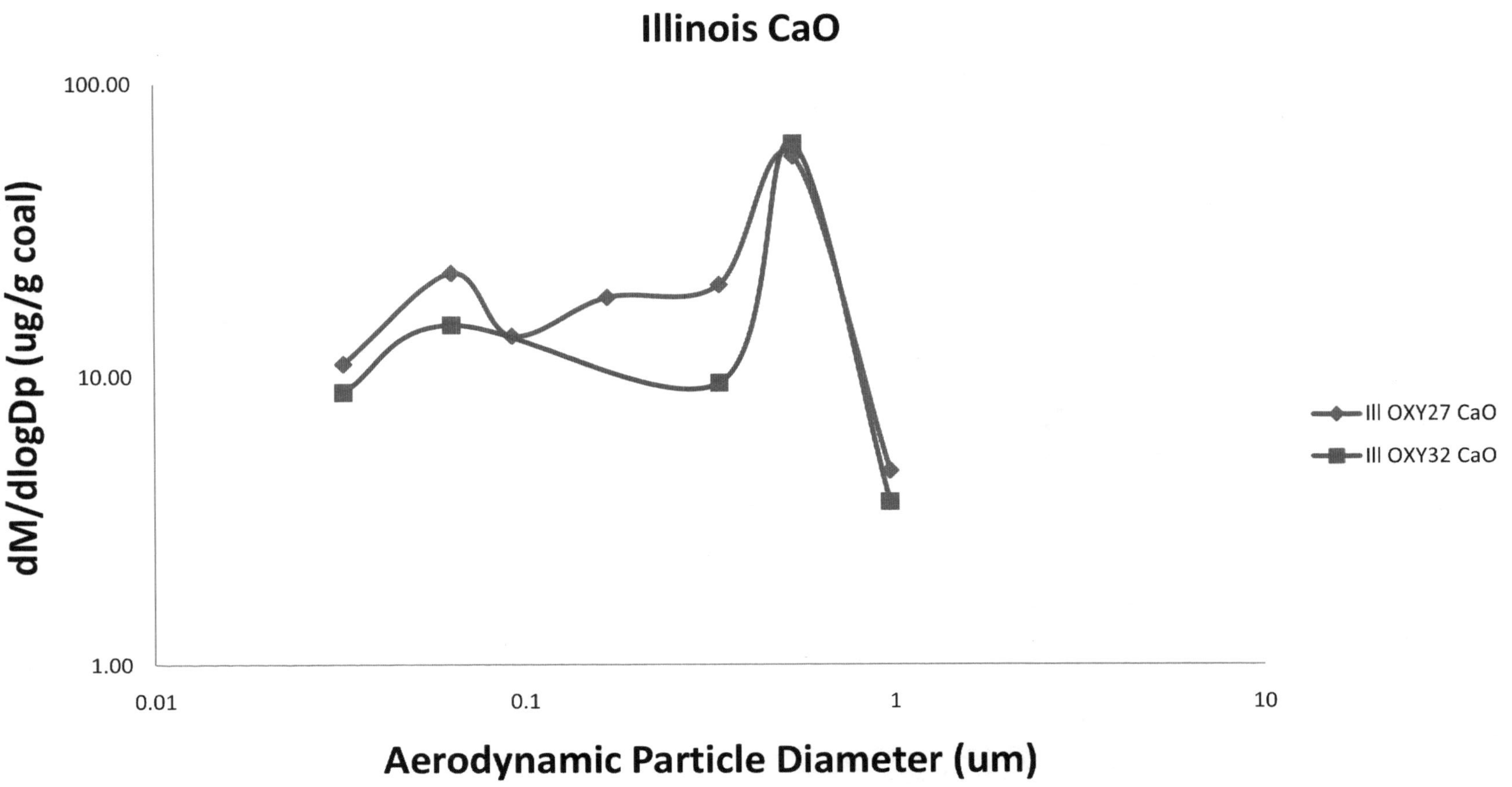
Illinois CaO
100.00
10.00
1.00
0.01
0.1
1
10
dM/dlogDp (ug/g coal)
Aerodynamic Particle Diameter (um)
Ill OXY27 CaO
Ill OXY32 CaO

# Results

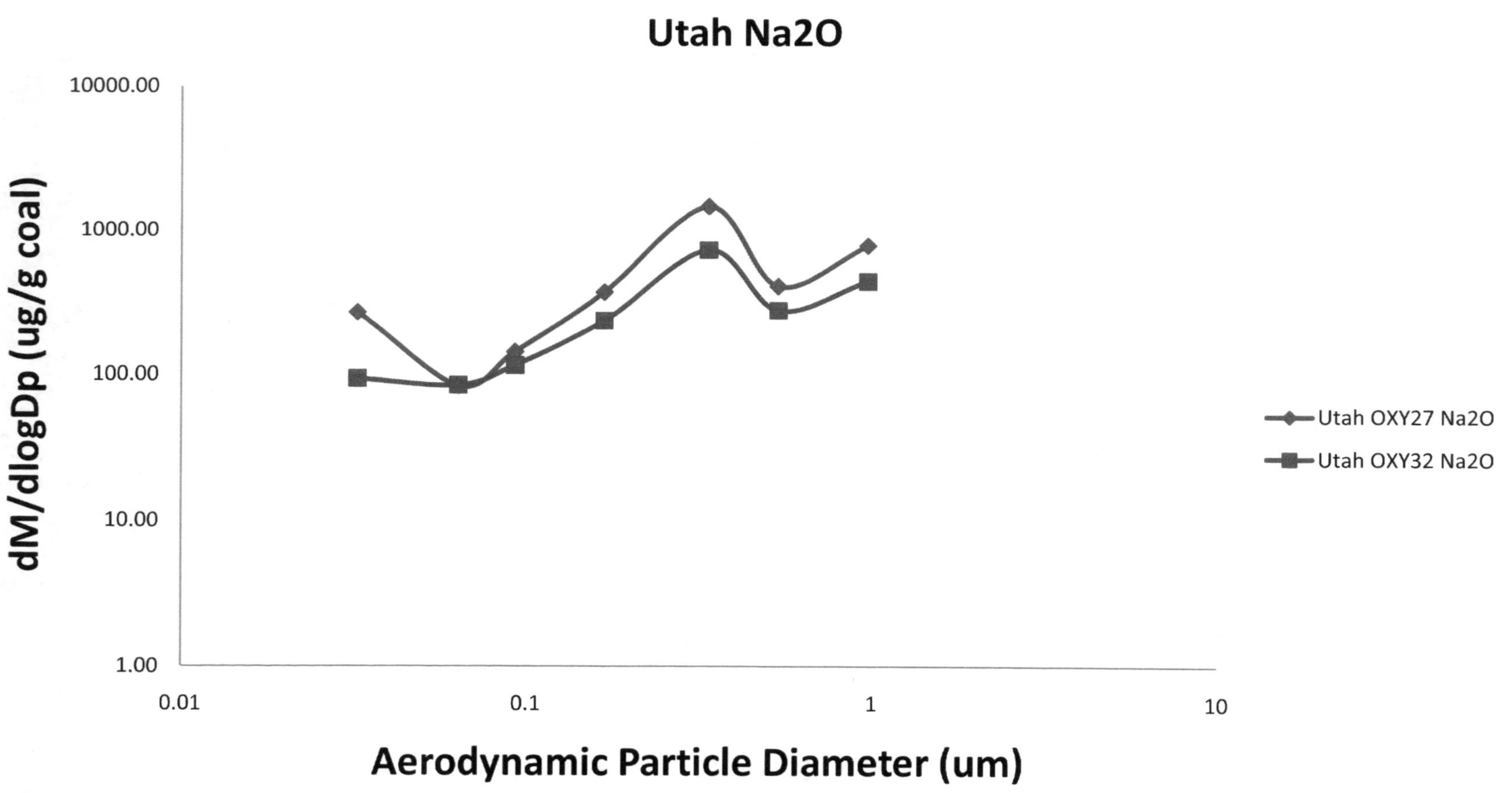
Utah Na2O
dM/dlogDp (ug/g coal)
10000.00
1000.00
100.00
10.00
1.00
0.01
0.1
1
10
Aerodynamic Particle Diameter (um)
Utah OXY27 Na2O
Utah OXY32 Na2O

# Results

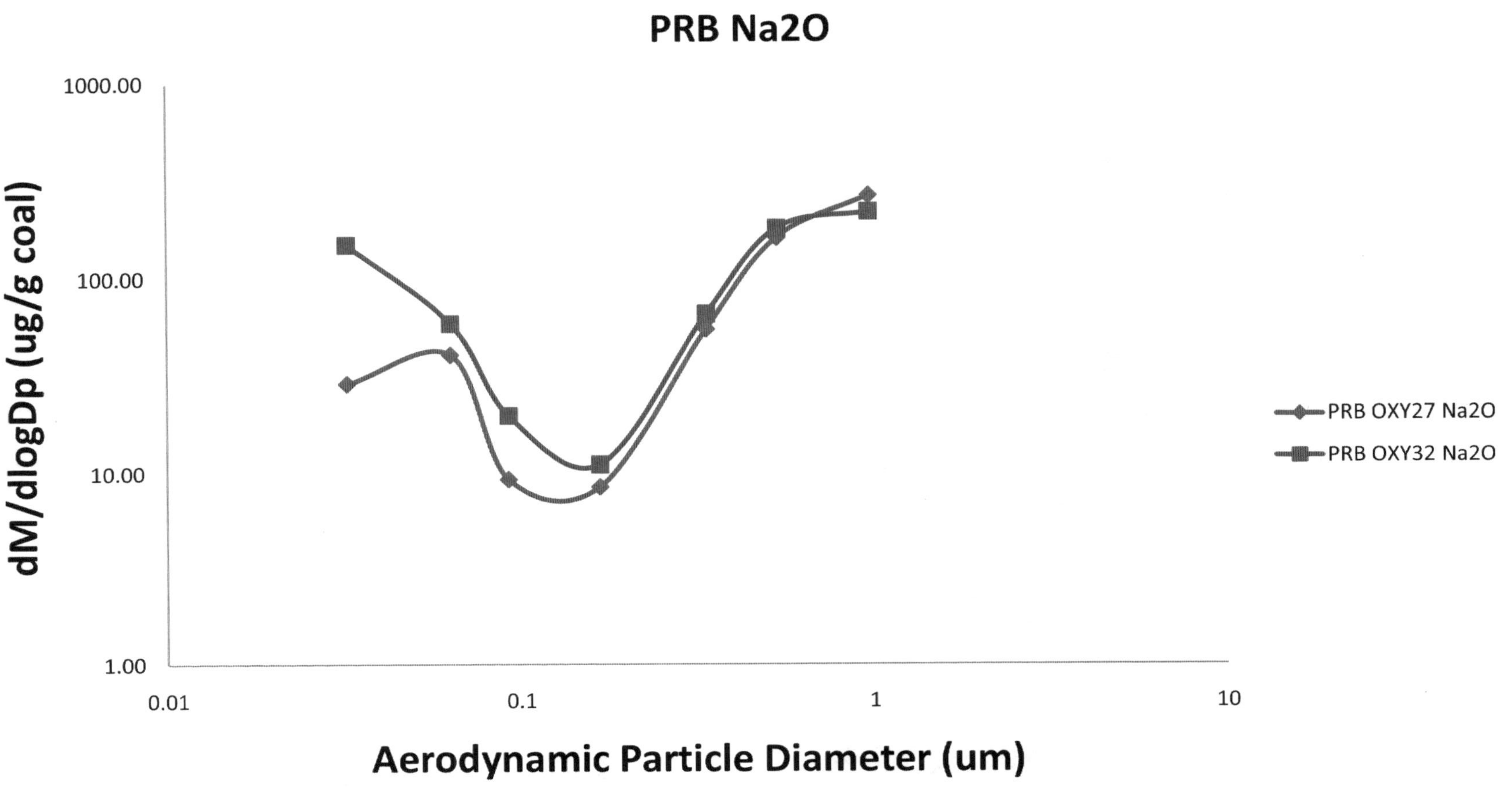
PRB Na2O
1000.00
100.00
10.00
1.00
dM/dlogDp (ug/g coal)
0.01
0.1
1
10
Aerodynamic Particle Diameter (um)
PRB OXY27 Na2O
PRB OXY32 Na2O

# Results

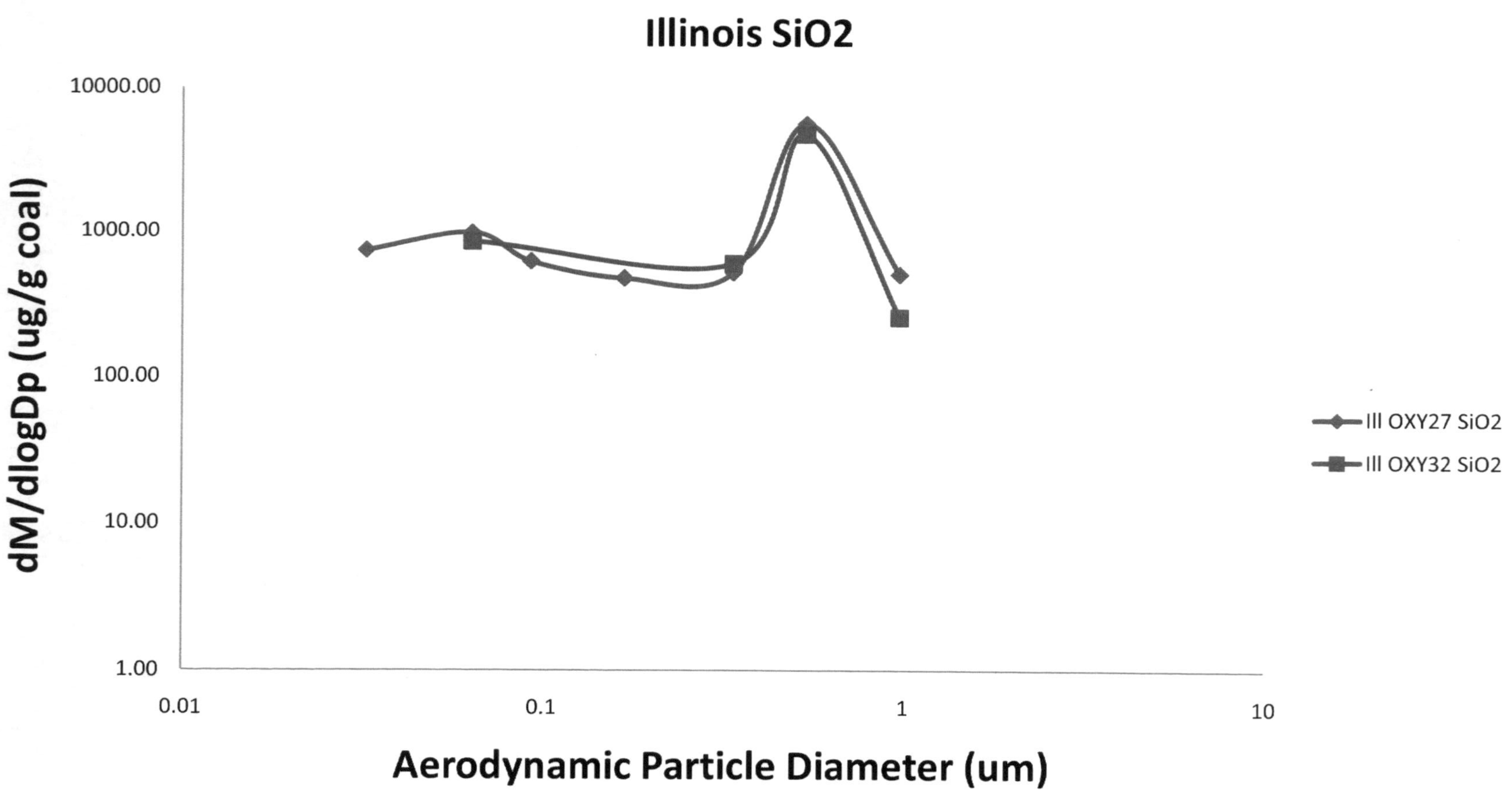
Illinois SiO2
dM/dlogDp (ug/g coal)
10000.00
1000.00
100.00
10.00
1.00
0.01
0.1
1
10
Aerodynamic Particle Diameter (um)
Ill OXY27 SiO2
Ill OXY32 SiO2

# Results

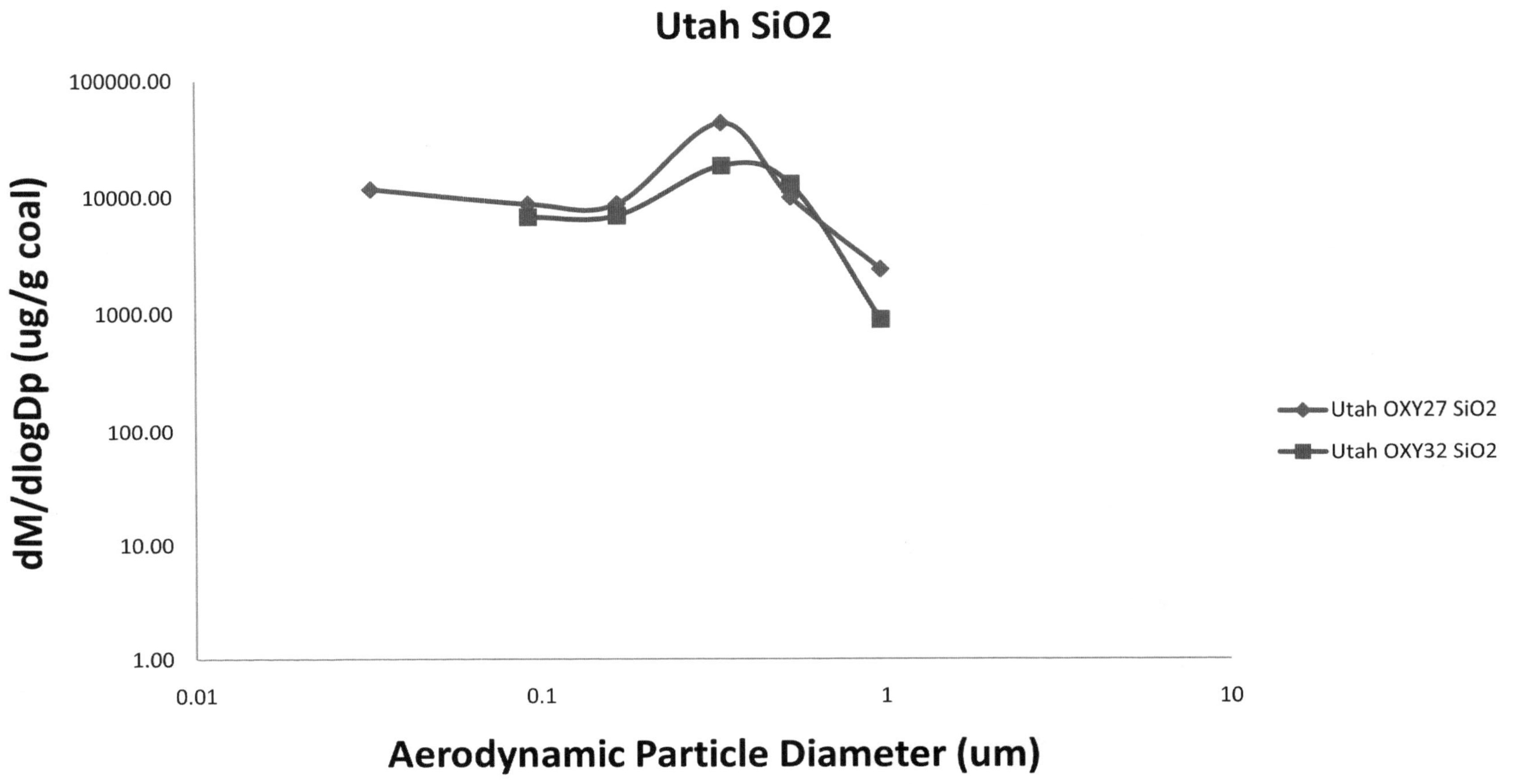
Utah SiO2
100000.00
10000.00
1000.00
100.00
10.00
1.00
0.01
0.1
1
10
dM/dlogDp (ug/g coal)
Aerodynamic Particle Diameter (um)
Utah OXY27 SiO2
Utah OXY32 SiO2

# Results

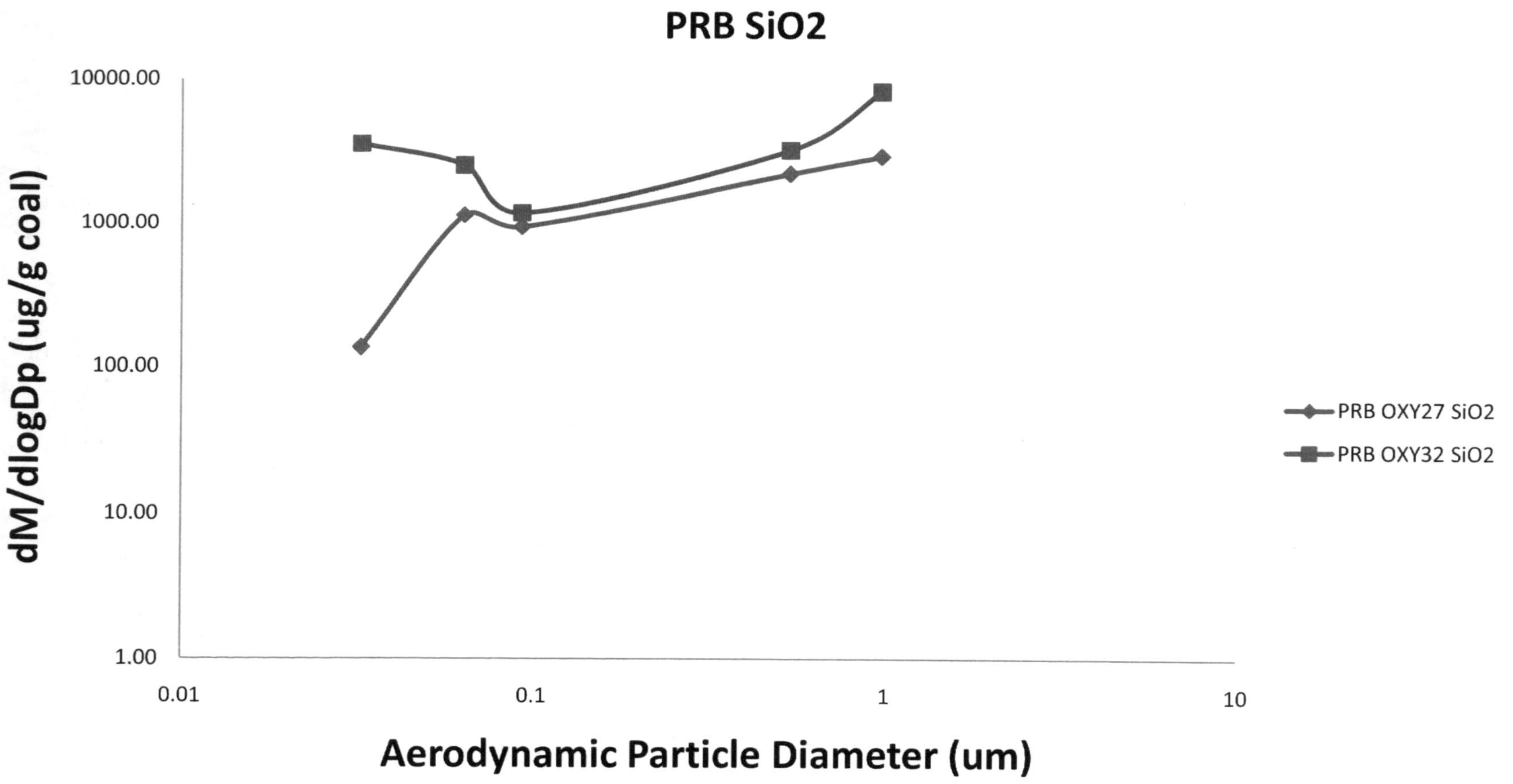
PRB SiO2
10000.00
1000.00
100.00
10.00
1.00
dM/dlogDp (ug/g coal)
0.01
0.1
1
10
Aerodynamic Particle Diameter (um)
PRB OXY27 SiO2
PRB OXY32 SiO2

# Results

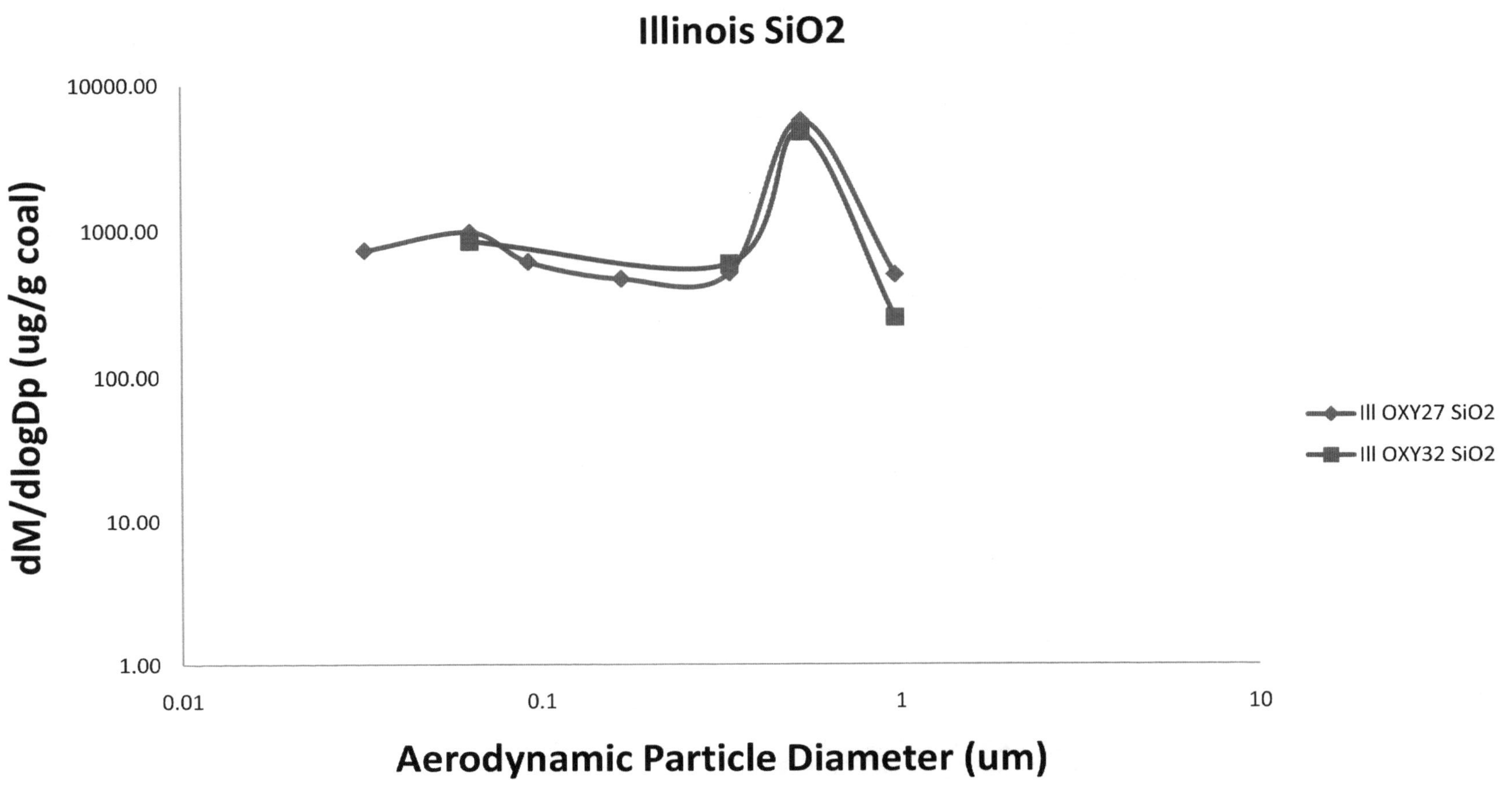
Illinois SiO2
10000.00
1000.00
100.00
10.00
1.00
dM/dlogDp (ug/g coal)
0.01
0.1
1
10
Aerodynamic Particle Diameter (um)
Ill OXY27 SiO2
Ill OXY32 SiO2

# Results

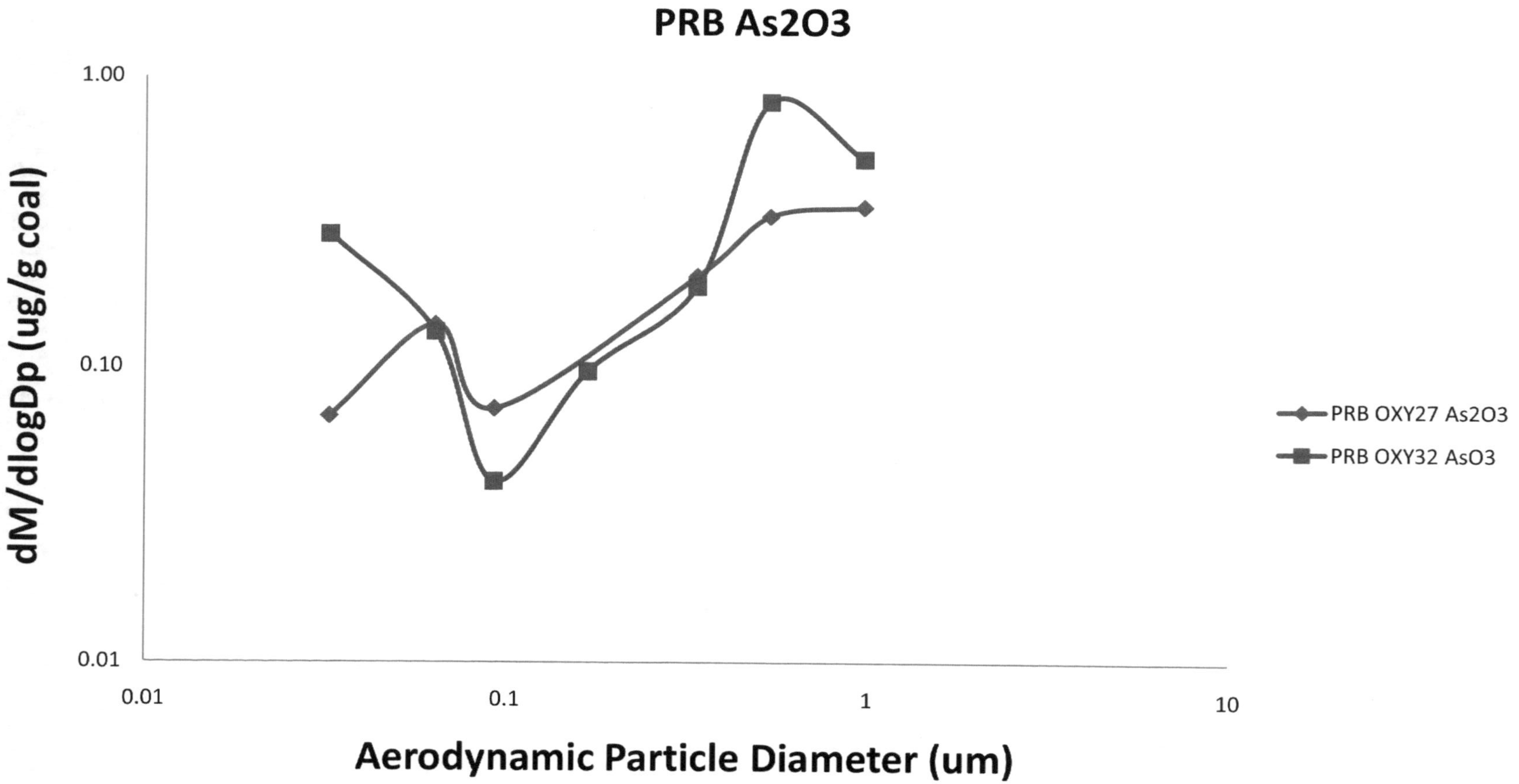
PRB As2O3
dM/dlogDp (ug/g coal)
1.00
0.10
0.01
0.01
0.1
1
10
Aerodynamic Particle Diameter (um)
PRB OXY27 As2O3
PRB OXY32 AsO3

# Results

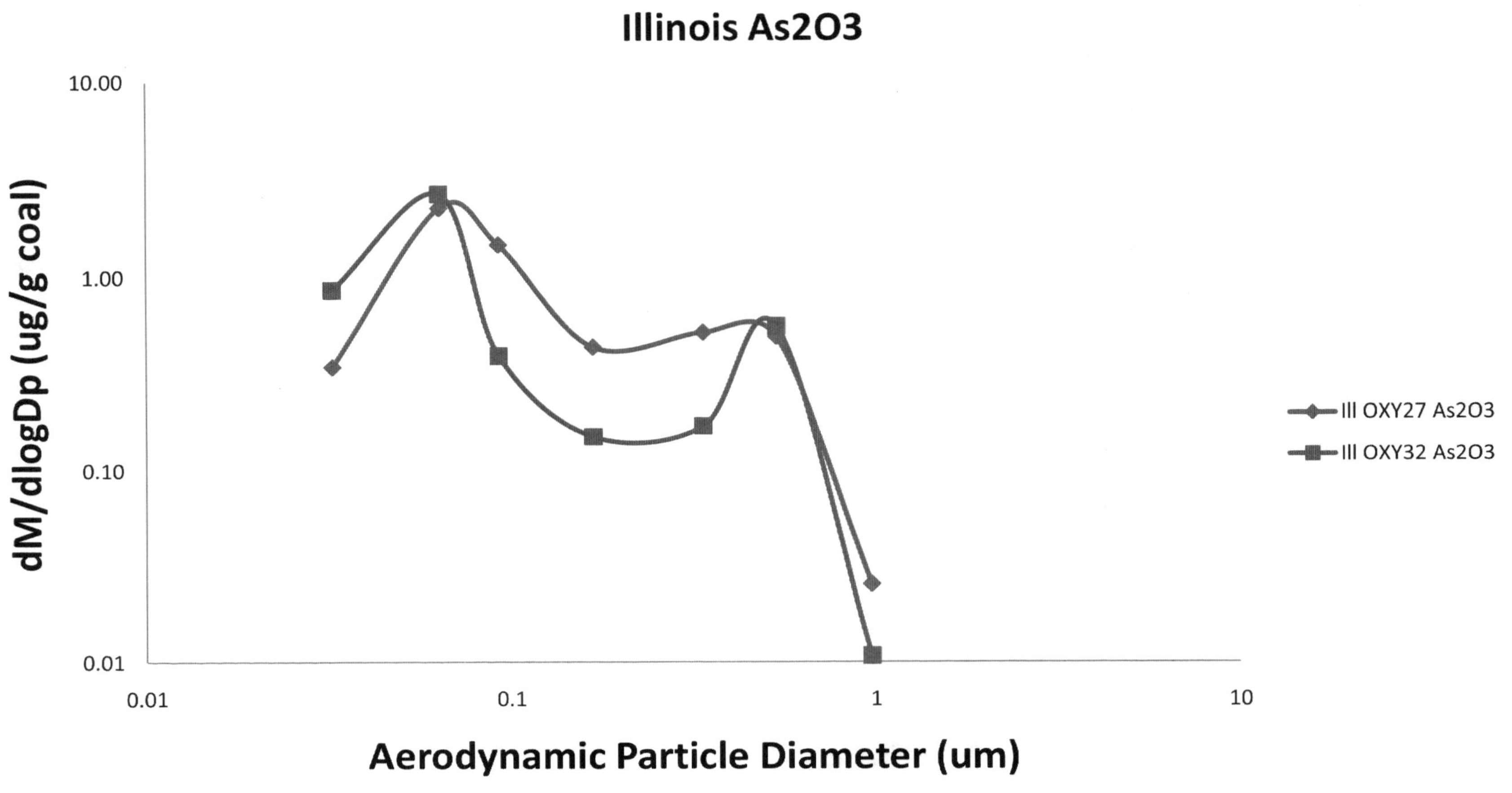
Illinois As2O3
10.00
1.00
0.10
0.01
dM/dlogDp (ug/g coal)
0.01
0.1
1
10
Aerodynamic Particle Diameter (um)
Ill OXY27 As2O3
Ill OXY32 As2O3

# Results

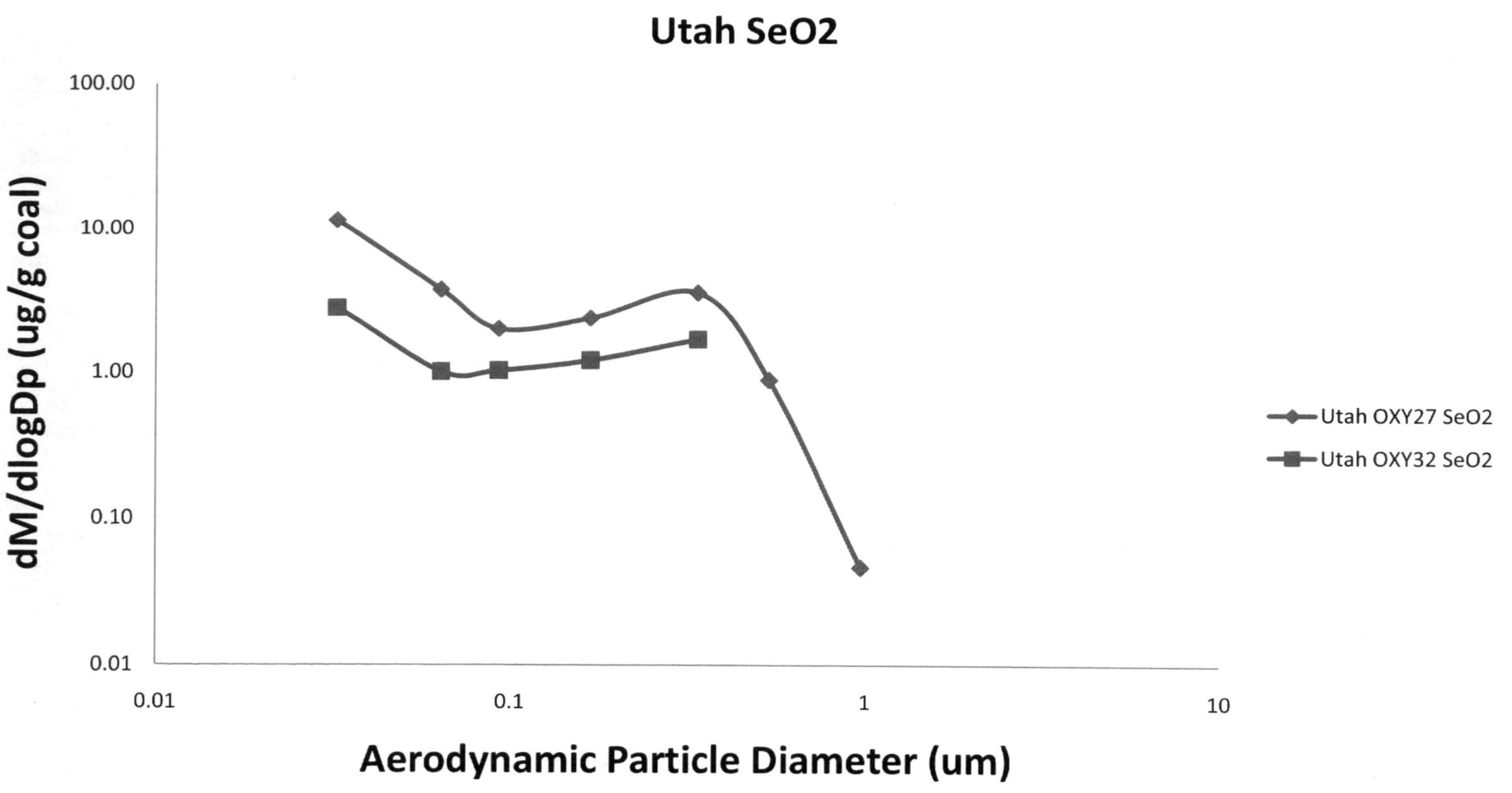
Utah SeO2
dM/dlogDp (ug/g coal)
100.00
10.00
1.00
0.10
0.01
0.01
0.1
1
10
Aerodynamic Particle Diameter (um)
Utah OXY27 SeO2
Utah OXY32 SeO2

# Results

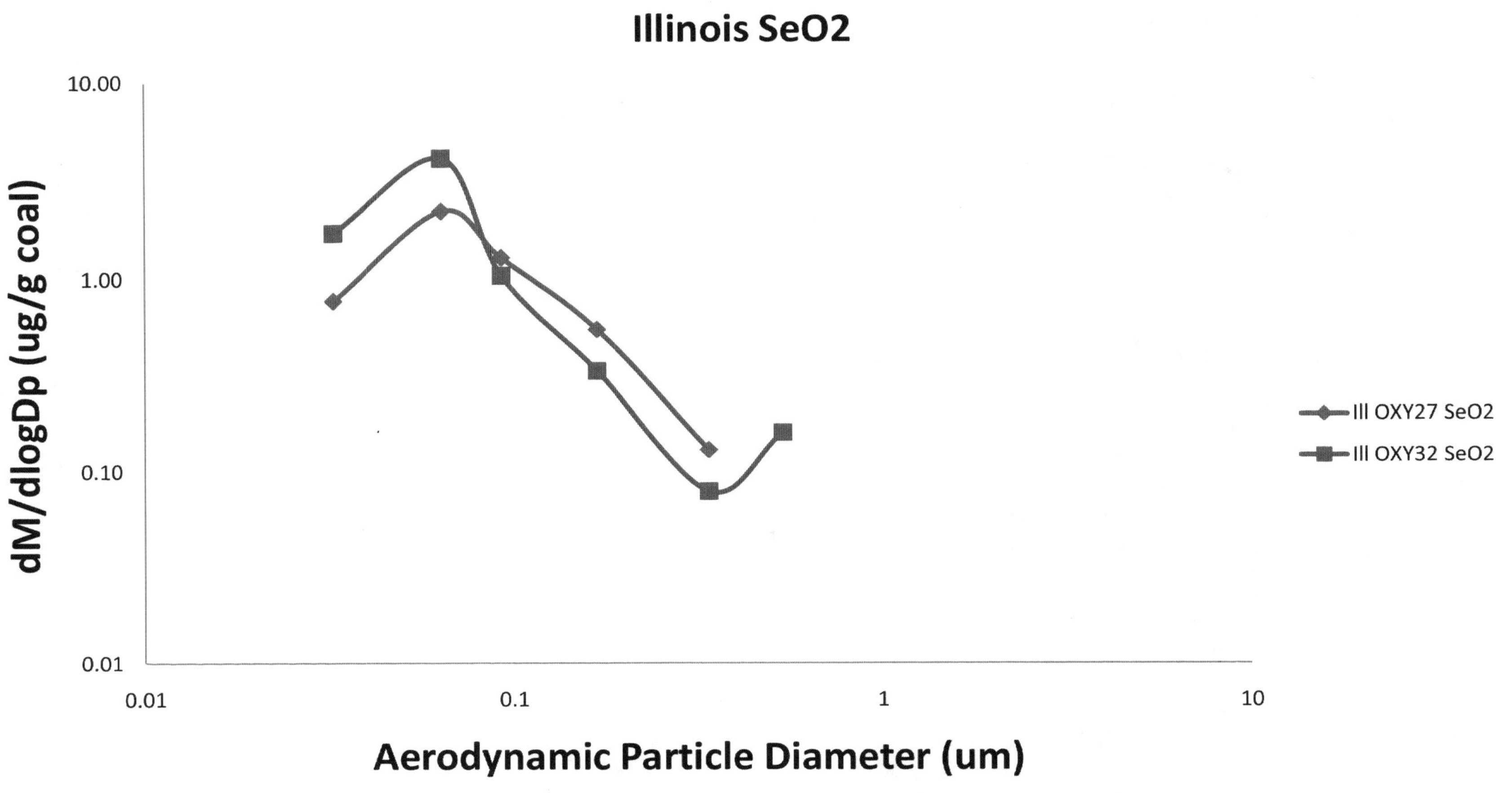
Illinois SeO2
dM/dlogDp (ug/g coal)
10.00
1.00
0.10
0.01
0.01
0.1
1
10
Aerodynamic Particle Diameter (um)
Ill OXY27 SeO2
Ill OXY32 SeO2

# Conclusions

- Fine fragmentation mode (~1 um) is not affected by small changes in $PO_2$ at conventional temperatures.
- Coal composition still plays the dominant role in aerosol chemistry.
- Only differences are noticed in the ultra-fine region <100nm. This is the condensed vapor phase mode and is likely due to a hotter flame temperature yielding increased vaporization.

# Acknowledgements


- Financial support from the Department of Energy under Awards DE-FC26-06NT42808 and DE-FC08-NT0005015
- David Wagner, Ryan Okerlund, Brian Nelson, Rafael Erickson, and Colby Ashcroft Institute for Clean and Secure Energy, University of Utah
- Diego Fernandez and his team of analytical chemists in the Department of Geology and Geophysics, University of Utah

# (487b) Laboratory-Scale Study of Fine Particle Formation and Mercury Speciation During Oxy-Coal Combustion

S. Michael Daukoru[1], Lei Zhang[2], Sarah Torkamani[1], Wei-Ning Wang[1], Shuxiao Wang[2], Jiming Hao[2] and **Pratim Biswas**[1], (1)Department of Energy, Environmental and Chemical Engineering, Washington University in St. Louis, St. Louis, MO, (2)Department of Environmental Science and Engineering, Tsinghua University, Beijing, China

Coal combustion is the largest single contributor to global anthropogenic $CO_2$ emissions, contributing ~50% of electricity generation in the United States and 42% of total $CO_2$ emissions. Several options exist for capturing $CO_2$, and oxy-coal combustion is a means for increasing the concentration of $CO_2$ in flue gas and minimizing parasitic losses in supplying $CO_2$ ready for geological sequestration or conversion to useful products. Ongoing work at the Aerosol & Air Quality Research Laboratory (AAQRL) at Washington University in St. Louis (WUStL) has focused on the oxy-coal combustion modality as a basis for producing flue gas with $CO_2$ of sufficiently high purity for direct conversion and sequestration applications, without expensive gas separation.

Laboratory-scale furnaces are used for various aspects of oxy-coal combustion, including a drop-tube furnace for studying the effect of flue-gas recycle on fine particle formation and mercury speciation, and a furnace aerosol reactor for synthesizing nanostructured catalysts for $CO_2$ photoreduction. The effect of gas composition on the sub-oxide formation in oxy-coal combustion systems will be discussed. As flue gas recycle may potentially impact the volatilization rates, this is also examined for submicrometer particle formation in contrast to systems with no recycle. The presentation will also discuss the resultant size distribution of particles and mercury speciation for a variety of coal seams from China and the United States in oxy-coal combustion systems, and contrast to those from conventional air combustion systems. Results indicate that the combustion in an oxygen-carbon dioxide environment suppresses the volatilization rates, resulting in a decrease in the fine particle fractions.

See more of this Session: Oxycombustion of Coal - Need, Opportunities, and Challenges II
See more of this Group/Topical: Topical E: High Temperature Environmentally Sustainable Energy Processes (sessions joint with the Environmental Division)

# (487c) Particulate Formation From Pulverized Coal Under Oxy-Fuel Combustion Conditions

**Yunlu Jia** and JoAnn Slama Lighty, Department of Chemical Engineering, University of Utah, Institute for Clean and Secure Energy, Salt Lake City, UT

Oxy-fuel coal combustion has been recognized as a promising technology for pulverized coal-fired power plants. This research is investigating particulate formation and morphology of ash particulate produced from oxy-fuel coal combustion in a high-temperature drop tube furnace. In this study, ash particulate has been studied as function of temperature, coal type and gas-phase conditions, namely carbon dioxide versus nitrogen. The ash particle size distributions were obtained by a Scanning Mobility Particle Sizer (SMPS) and Aerodynamic Particle Sizer (APS), which allows for measurement of particles in size ranges between 14.3 nm to 20 microns. A Berner Low Pressure Impactor (BLPI) was also used to collect particles on eleven stages to perform morphology and element analysis by Scanning Electron Microscopy (SEM) and Energy Dispersive Spectroscopy (EDS). A tri-model particle size distribution including an ultrafine mode, a submicron mode and a supermicron mode was found. Furthermore, the results showed that in the presence of carbon dioxide, less ultrafine particles were formed, as compared to nitrogen. These results support the hypothesis that the refractory oxides are governed by the equation: MO(s) + CO(g) $\leftrightarrow$ CO2(g) + M(g).

See more of this Session: Oxycombustion of Coal - Need, Opportunities, and Challenges II
See more of this Group/Topical: Topical E: High Temperature Environmentally Sustainable Energy Processes (sessions joint with the Environmental Division)

# ASH FORMATION AND DEPOSITION DURING OXY-COAL COMBUSTION

*Dunxi Yu, William J. Morris, Raphael Erickson, Jost O. L. Wendt*
*Department of Chemical Engineering, University of Utah, Salt Lake City, UT*

*Andrew Fry, Constance L. Senior*
*Reaction Engineering International, Salt Lake City, UT*

## Introduction

Coal combustion for power generation is a major source of $CO_2$ and several CCS (carbon capture and storage) technologies are being actively investigated [1]. Oxy-fuel combustion is an attractive option for coal-based power generation with $CO_2$ capture. Compared with conventional air combustion, oxy-coal combustion may differ in many aspects such as higher $O_2$ content in the feed gas, higher concentrations of $CO_2$ and $H_2O$ in the flue gas, lower volumetric gas flow and longer residence time of flue gas and particles in the furnace [1, 2]. Such differences have been shown to affect ignition, flame stability, char combustion, heat transfer, gas emissions, and possibly ash deposition [3].

Ash deposition is particularly a concern in oxy-fuel combustion because it may be affected by a number of parameters, such as changed temperature profiles, increased ash loading in flue gas [4], increased capture of sulfur by the ash [5, 6], and increased carbonation of fly ash [7]. Although some studies have been carried out, there's still no consensus on ash deposition behavior [3]. The present work aims to investigate ash formation, which is a critical step in the development of deposits, and ash deposition during oxy-coal combustion with once-through $CO_2$.

## Experimental

Two coals of different ranks were tested, including a PRB subbituminous coal and an Illinois bituminous coal. Their properties and some combustion parameters are shown in Table 1. The PRB coal ash has higher contents of CaO and MgO while the Illinois coal ash contains more $Fe_2O_3$. Three combustion scenarios were selected: (1) air combustion; (2) oxy-fuel combustion with $27\%O_2/73\%CO_2$; (3) oxy-fuel combustion with $32\%O_2/68\%CO_2$. They are hereinafter donated as AIR, OXY27 and OXY32, respectively. The OXY27 case was designed to match the gas radiation fluxes in air combustion while the OXY32 case matched the adiabatic flame temperature. Once-through $CO_2$ was used in oxy-fuel cases to simulate cleaned flue gas recycle with all contaminants and water removed. The excess oxygen in the flue gas was fixed at 3%, similar to practical pulverized coal combustion.

**Table 1.** Coal properties and combustion parameters

| | BLACK THUNDER PRB | ILLINOIS |
|---|---|---|
| Proximate analysis (ar, wt%) | | |
| Moisture | 23.69 | 9.65 |
| Volatile Matter | 33.36 | 36.78 |
| Fixed Carbon | 38.01 | 45.58 |
| Ash | 4.94 | 7.99 |
| Ultimate analysis (ar, wt%) | | |
| C | 53.72 | 64.67 |
| H | 6.22 | 5.59 |
| N | 0.78 | 1.12 |
| S | 0.23 | 3.98 |
| O (diff) | 34.11 | 16.65 |
| Ash analysis (wt%) | | |
| $Al_2O_3$ | 14.78 | 17.66 |
| CaO | 22.19 | 1.87 |
| $Fe_2O_3$ | 5.2 | 14.57 |
| MgO | 5.17 | 0.98 |
| MnO | 0.01 | 0.02 |
| $P_2O_5$ | 1.07 | 0.11 |
| $K_2O$ | 0.35 | 2.26 |
| $SiO_2$ | 30.46 | 49.28 |
| $Na_2O$ | 1.94 | 1.51 |
| $SO_3$ | 8.83 | 2.22 |
| $TiO_2$ | 1.3 | 0.85 |
| Combustion parameters | | |
| Coal flow rate (kg/h) | 6.26 | 4.89 |
| Coal firing rate (kW) | 36.64 | |

All combustion tests were performed on a down-fired oxy-fuel furnace with a designed capacity of 100kW (Fig. 1), which was designed to simulate practical units and allowed systematic variations of inlet oxidant mixtures from air to $O_2/CO_2$. Nine groups of opposite ports, used for observation and sampling, were lined along the vertical section of the furnace. Detailed configuration information is available elsewhere [8].

Both bulk ash and size-segregated ash were sampled through the lowest port on the vertical section of the furnace (Port 9). The system used to collect bulk ash consisted of a water-cooled probe with a 90 mm filter housing, where ash was deposited on the filter paper for subsequent examination. No dilution was used. Flow through the probe was induced using a vacuum pump and control of the flow rate was adjusted using a rotameter. The size-segregated ash sampling system included another water-cooled, nitrogen-quenched isokinetic sampling probe, a pre-cyclone and a Berner Low

Pressure Impactor (BLPI). The BLPI consisted of 11 stages with corresponding 50% aerodynamic cutoff diameters as 15.7, 7.33, 3.77, 1.98, 0.973, 0.535, 0.337, 0.168, 0.0926, 0.0636 and 0.0324 μm. The total flow rate was kept constant due to a critical orifice within the impactor at approximately 25.4 L/min and verified using a Gillibrator bubble flow meter. Cellulose acetate membranes coated with high purity Apiezon grease were used to collect particles for composition analysis while aluminum foils were used to obtain PSDs. For the Illinois coal, the BLPI had to be used in two separate configurations to obtain a complete PSD due to the high loading of supermicron size particles. Details on these configurations are available elsewhere [9]. For the PRB coal, a complete PSD could be obtained in the same run but the sampling time had to be greatly reduced.

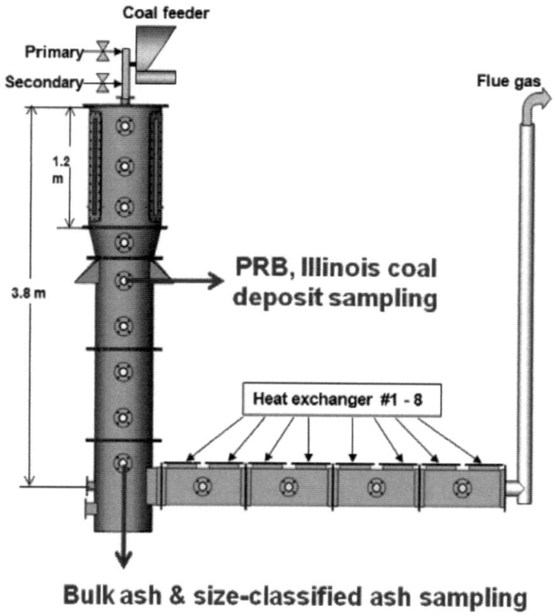


**Figure 1.** Experimental facility

Deposits were collected on un-cooled ceramics that was attached to a water-cooled deposition probe. Deposition tests were carried out at Port 5, just below the ignition zone of the furnace. The exposure time for the PRB coal tests was about 4 hours while that for the Illinois coal tests was about 5 hours. Due to changes in combustion conditions, gas temperatures at the deposition site (Port 5) were different. For the PRB coal, gas temperatures measured in cases AIR, OXY27 and OXY32 were 1298 K, 1313 K and 1316 K, respectively. For the Illinois coal, gas temperatures measured in cases AIR, OXY27 and OXY32 were 1311 K, 1339 K and 1361 K, respectively. The data show that, at the deposition site, gas temperatures in oxy-fuel combustion are generally higher than in air combustion. For the purpose of composition analysis, some deposits were scraped from the ceramics and ground to powder samples so that heterogeneity could be reduced.

All bulk ash, size-segregated ash and deposit samples were subject to composition analysis by a FEI NanoNova Scanning Electron Microscope (SEM) coupled with an Energy Dispersive X-ray Spectrometer (EDS). Ten common oxides were considered, i.e., $Al_2O_3$, $SiO_2$, $CaO$, $Fe_2O_3$, $Na_2O$, $K_2O$, $MgO$, $SO_3$, $P_2O_5$, and $TiO_2$.

## Results

### *Data on ash formation*

Ash particles are generated from coal mineral matter and their formation is affected not only by mineral matter properties but also by the temperature and gas composition. Switching from air-firing to oxy-firing leads to changes in combustion atmosphere and possible changes in temperatures. Therefore, ash properties and formation processes may be different during oxy-coal combustion.

The composition data of the bulk ashes collected under different combustion conditions are compared in Fig. 2. It's apparent that ash composition is coal-dependent. The PRB coal ashes consist primarily of CaO, $SiO_2$, and $Al_2O_3$. There are also appreciable concentrations of $Fe_2O_3$, MgO, $SO_3$ and $Na_2O$. In contrast, the Illinois coal ashes are dominated by $Al_2O_3$, $SiO_2$ and $Fe_2O_3$. For each coal, however, the ash composition is very similar between oxy-fuel and air combustion. This suggests that oxy-firing has insignificant effects on ash formation from the point of view of changes in oxide constituents.

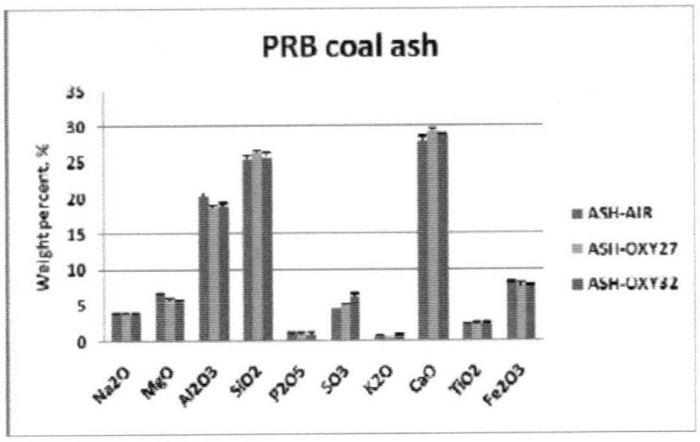


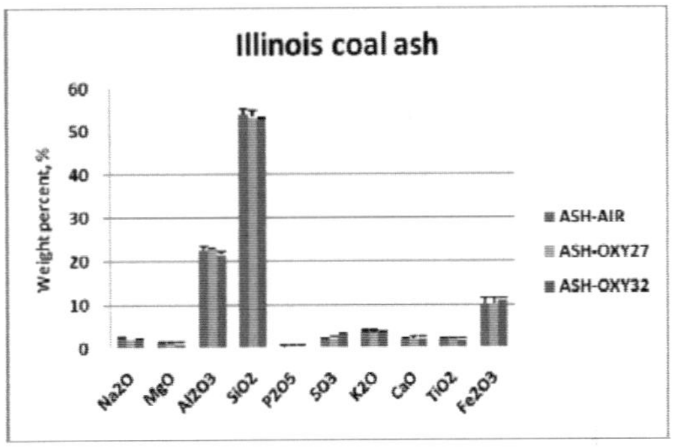


**Figure 2.** Comparison of bulk ash composition

Fig. 3 depicts size-segregated ash particle size distributions (PSDs). Generally, there are no significant differences between air- and oxy-firing cases, except the PRB coal, where there seems to be some differences between ash PSDs >2 μm. It indicates that oxy-firing may not have appreciable effects on the concentration of particles <10 μm. But it should be noted that the flue gas volume is different for different combustion cases, e.g. the OXY27 flue gas volume is about 75% of that of the AIR case while the OXY32 flue gas volume about 63% of that of the AIR case. As a result, the total amount of particles <10 μm may be affected by oxy-coal combustion.

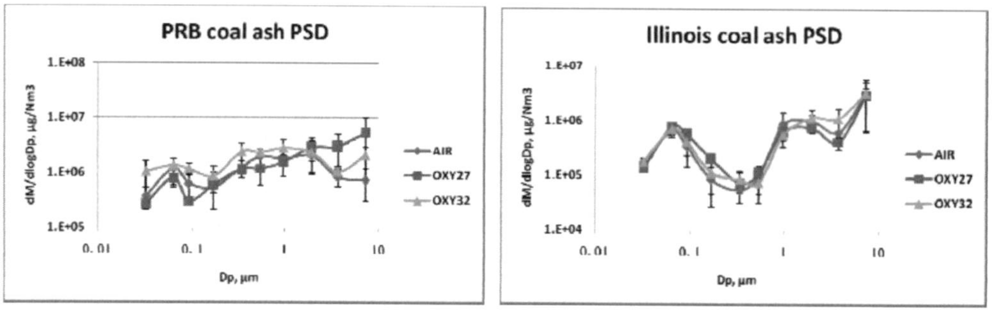


**Figure 3.** Ash aerosol size distributions

The size-segregated ash samples were also subjected to chemical composition analysis. But due to larger uncertainties of very small particles, only data of particles >0.2 μm are reported. Fig. 4 shows typical size distributions of CaO, MgO, $Fe_2O_3$ and $SO_3$ in the PRB coal size-segregated ashes. The profiles generally follow the ash PSDs as shown in Fig. 3. But since the oxy-firing cases have lower flue gas volume, the total amount of some oxides in particles <10 μm may be affected as well.

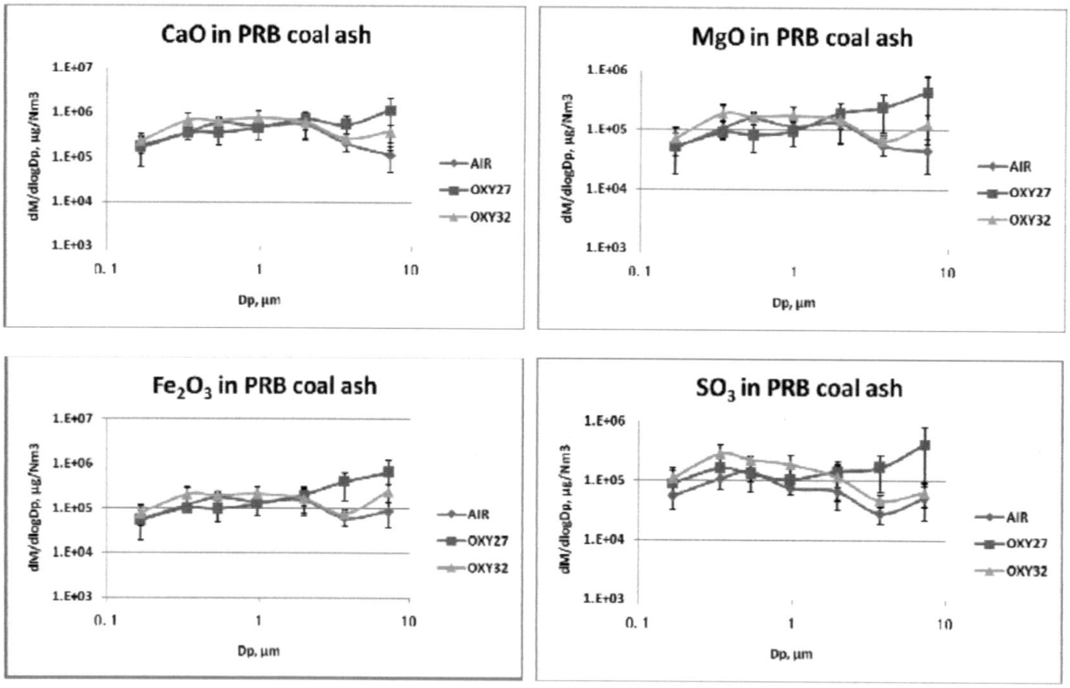


**Figure 4.** Oxide size distributions

***Data on ash deposition***

Figure 5 shows ceramics loaded with deposits. The PRB coal deposits are thin and tight (Fig. 5a) while the Illinois coal deposits are thicker and looser (Fig. 5b). The colors of the deposits are also different between the two coals. These differences suggest that ash deposition is coal-dependent. In terms of color and morphology, the deposits do not vary significantly between oxy-fuel and air combustion. However, the thickness of the oxy-firing deposits is thicker than that of the air-firing deposits. This is especially obvious for the Illinois coal. Such difference demonstrates that oxy-firing may have notable effects on deposition rate.

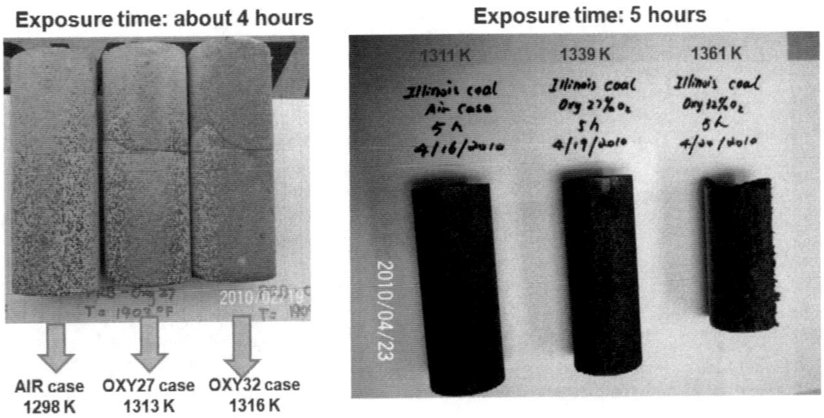


(a) PRB coal (b) Illinois coal

**Figure 5.** Ceramics loaded with deposits

The composition of the deposits collected under different combustion conditions is shown in Fig. 6. Compared with the AIR case, the oxy-firing deposits of the PRB coal contain higher concentrations of CaO, MgO and $Al_2O_3$, but lower concentrations of $SO_3$ and $Fe_2O_3$. For the Illinois coal, the oxy-firing deposits are enriched in $Al_2O_3$ and $SiO_2$ but depleted in $Fe_2O_3$. These differences suggest that oxy-firing also has influence on deposit composition.

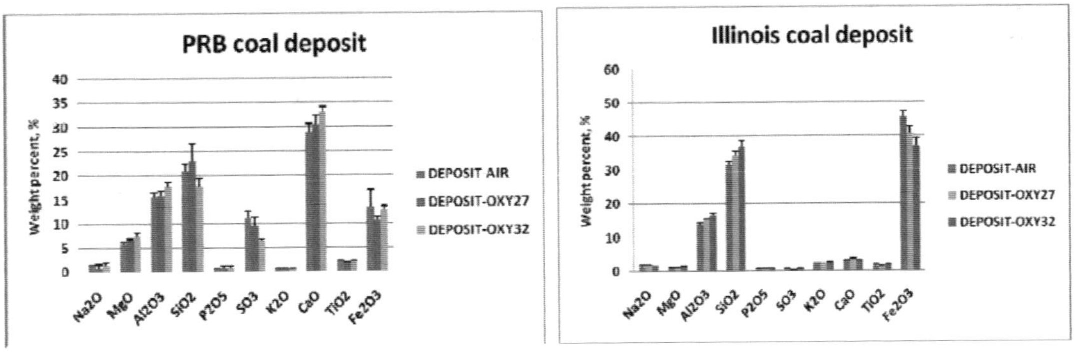


**Figure 6.** Deposit composition

For the PRB coal, the deposits have higher concentrations of $Fe_2O_3$ and CaO compared to the ashes (Fig. 2), regardless of combustion scenarios. In the case of the Illinois coal, the $Fe_2O_3$ content in the deposits is almost three times higher than that in the ashes (Fig. 2).

## Discussion and conclusions

Experimental data show that oxy-firing has insignificant effects on the oxide composition of the bulk ashes, the concentrations of particles <10 μm and the major oxide constituents in them. This may give one an impression that oxy-firing should not have significant influence on ash deposition. However, Fig. 5 suggests that oxy-firing may accelerate ash deposition under combustion conditions studied. In addition, the chemical composition of the oxy-firing deposits is different from that of the air-firing deposits, as shown in Fig. 6. These results show that oxy-firing does have appreciable effects on ash deposition. Such seeming inconsistency suggests that more related factors should be considered as well in understanding ash deposition during oxy-fuel combustion.

1. Although the coal ashes do not differ much from the point of view of oxide composition between air and oxy-fuel combustion, it may not be the same case in light of species composition. For example, due to the higher $O_2$ partial pressure, oxy-firing is expected to favor the formation of $Fe_3O_4$ and $Fe_2O_3$ rather than FeO [10], which is an important deposition species. Because of changes in gas composition, sulfation and carbonation of ash particles may be enhanced under oxy-fuel combustion [11]. The effects of these changes on ash deposition have not been well addressed. It suggests that only the knowledge of the oxide composition may not be enough, and chemical speciation may be necessary.

2. The difference in flue gas volume between oxy-fuel and air combustion must be considered because of its important effects on furnace aerodynamics. The flue gas volume for the OXY27 case is about 75% of that for the AIR case while the flue gas volume for the OXY32 case is reduced to about 63% of that for the AIR case. On the one hand, the reduced gas volume in the oxy-fuel cases leads to longer residence time and lower velocities of particles in the furnace. On the other hand, it also results in higher ash loading in the oxy-fuel flue gas, as the air and oxy-fuel cases used the same firing rate. These changes are expected to affect ash deposition to various extents.

3. Gas temperature can have important influence on ash deposition. Generally, higher gas temperatures can promote ash deposition by increasing melting of the fuel ash. In this study deposition tests were carried out at the same location. Gas temperatures in oxy-fuel combustion were a dozen or even several dozen K higher than those in air combustion. This may contribute to the enhanced ash deposition in oxy-fuel combustion.

## Acknowledgements

This work is supported by the US Department of Energy under Award Numbers DE-NT0005015 and DE-FC26-06NT42808. The funding from the National Natural Science Foundation of China under Award Number 50720145604 is also acknowledged. The assistance of David Wagner, Ryan Okerlund, Brian Nelson, and Colby Ashcroft of the University of Utah is greatly appreciated.

## *References*


[1] Wall, T. F. (2007), "Combustion processes for carbon capture," *Proceedings of the Combustion Institute*, 31(1): 31-47.

[2] Buhre, B. J. P., Elliott, L. K., Sheng, C. D., et al. (2005), "Oxy-fuel combustion technology for coal-fired power generation," *Progress in Energy and Combustion Science*, 31(4): 283-307.

[3] Toftegaard, M. B., Brix, J., Jensen, P. A., et al. (2010), "Oxy-fuel combustion of solid fuels," *Progress in Energy and Combustion Science*, 36(5): 581-625.

[4] Wall, T., Liu, Y., Spero, C., et al. (2009), "An overview on oxyfuel coal combustion--State of the art research and technology development," *Chemical Engineering Research and Design*, 87(8): 1003-1016.

[5] Liu, H., Okazaki, K. (2003), "Simultaneous easy CO2 recovery and drastic reduction of SOx and NOx in O2/CO2 coal combustion with heat recirculation[small star, filled]," *Fuel*, 82(11): 1427-1436.

[6] Liu, H., Katagiri, S., Okazaki, K. (2001), "Drastic SOx Removal and Influences of Various Factors in O2/CO2 Pulverized Coal Combustion System," *Energy & Fuels*, 15(2): 403-412.

[7] Wang, C., Jia, L., Tan, Y., et al. (2008), "Carbonation of fly ash in oxy-fuel CFB combustion," *Fuel*, 87(7): 1108-1114.

[8] Zhang, J., Okerlund, R., Eddings, E. G., et al. (2007), "Ignition of axial pulverized coal jets in a specially designed 100 kW oxy-fuel combustor," in: *6th International Symposium on Coal Combustion*, Wuhan, Hubei, China.

[9] Seames, W. S., Wendt, J. O. L. (2000), "Partitioning of arsenic, selenium, and cadmium during the combustion of Pittsburgh and Illinois #6 coals in a self-sustained combustor," *Fuel Processing Technology*, 63(2-3): 179-196.

[10] Sheng, C., Li, Y. (2008), "Experimental study of ash formation during pulverized coal combustion in O2/CO2 mixtures," *Fuel*, 87(7): 1297-1305.

[11] Kull, R., Stein-Brzozowska, M., Scheffknecht, G., et al. (2009), "Corrosion of superheater materials under oxyfuel conditions," in: *1st Oxyfuel Combustion Conference*, Cottbus, Germany.

# (487e) Development of a Bench-Scale Oxy-Fuel Combustion System for the Study of Pollutant Behavior

Myrrha Andersen[1], William L. Roberts Roberts[1], Jost O. L. Wendt[2], Chun Wai Lee[3] and William P. Linak[3], (1)Department of Mechanical and Aerospace Engineering, North Carolina State University, Raleigh, NC, (2)Department of Chemical Engineering, University of Utah, Salt Lake City, UT, (3)Air Pollution Prevention and Control Division, National Risk Management Research Laboratory, U.S. Environmental Protection Agency, Research Triangle Park, NC

In July 2008, the U.S. EPA proposed to regulate the underground injection of carbon dioxide ($CO_2$) for the purpose of geologic sequestration (GS) under the Safe Drinking Water Act (SDWA). Elements of this proposed rule will build upon existing Underground Injection Control (UIC) regulations, with modifications to address the unique nature of $CO_2$ injection for GS. UIC regulations address technologies developed and refined by the oil and gas, and chemical manufacturing industries over the past several decades, and when applied to $CO_2$ GS, would regulate the siting, construction, operation, monitoring and testing, and closure of injection wells that could endanger the safety of underground sources of drinking water as required by SDWA. One concern is the possible presence of impurities in the $CO_2$ stream. These impurities could include residual quantities of hydrogen sulfide, sulfur and nitric oxides (or acids), mercury and other volatile metals, as well as organic and inorganic hazardous pollutants.

Oxygen enriched combustion (oxy-fuel) and specifically, oxy-coal combustion, is a developing, and potentially a strategically key, technology intended to accommodate direct $CO_2$ recovery for possible sequestration. Oxy-coal combustion is applicable for new as well as for retrofit application to existing coal-fired power plants. During oxy-coal combustion, combustion air is separated into its oxygen ($O_2$) and nitrogen ($N_2$) components and the coal is burned in a mixture of $O_2$ and recycled flue gas. By eliminating the $N_2$, ~80% of the volumetric flow through the utility plant is eliminated. The resulting effluent is composed primarily of $CO_2$, $H_2O$, and small amounts of residual $O_2$, $N_2$ (in leakage), and pollutant species ($SO_x$, $NO_x$, fly ash, metals, Hg, HCl, organics, etc.) which must be further processed before the $CO_2$ can be compressed, transported, and sequestered.

The state of the science behind oxy-fuel/oxy-coal combustion technology is far from mature, although a number of academic, governmental, and industrial research organizations are investigating various aspects of the technology. These ongoing efforts are focusing primarily on resolving technical and operational issues such as ignition and flame stability, fouling, slag behavior, corrosion, heat transfer, and efficiency. There is a significant lack of understanding regarding the environmental issues which must also be addressed before oxy-fuel/oxy-coal combustion can be adopted commercially. These include effects on pollutant formation, and how changes in pollutant behavior, concentrations, and the flue gas environment affect existing pollution control devices, and the characteristics of the various process and waste streams. Further, it is unknown how pollutant species will partition within compression and purification units (CPUs) designed to produce a supercritical $CO_2$ effluent for transportation and sequestration. Key questions include what purity of $CO_2$ is required to ensure safe long term GS. For oxy-coal combustion, specific questions include whether trace levels of sulfuric and nitric acids and other impurities can be compressed and sequestered with the $CO_2$, and what are the characteristics of the ash and especially the fly ash and volatile metals such as mercury and selenium (Hg, Se) that would normally penetrate the particle control devices as an air emission, but now may be compressed and sequestered with the $CO_2$. Other questions include the effect of a $CO_2$ environment on particle charging and electrostatic precipitator (ESP) operation, as well as similar concerns related to the operation of selective catalytic reduction (SCR) units and acid gas scrubbers.

To address some of these environmental issues, the U.S. EPA's National Risk Management Research Laboratory is modifying several in-house combustion facilities to conduct oxy-fuel experiments. This presentation will describe the design and construction of a small bench-scale (5 cm inside diameter by 1.5 m long) externally heated alumina reactor designed to burn natural gas and coal (~1-2 g/min) with mixtures of $O_2$ and $CO_2$. Initial experiments will examine one-pass $O_2$-$CO_2$ with natural gas, and evolve in the future to include coal and flue gas recirculation. The system is designed to examine realistic combustion stoichiometries and residence times. Measurements will examine the effects of combustion environment ($O_2$, $CO_2$, $H_2O$, fuel N, fuel S, and temperature) on $NO_x$ and $SO_x$ species (doped natural gas) and ash vaporization (coal). Trace metal behavior (including Hg and Se) are of particular interest. Measurements will include particle size distributions, extent of metal vaporization, and metal speciation. Variations in NO, $NO_2$, $N_2O$, and $SO_2$ emissions as well as efficiency (CO and ash carbon) will be examined and compared to air combustion.

This project is funded in part by the NCSU/EPA Cooperative Training Program in Environmental Science Research, Training Agreement CT833235-01-0 with North Carolina State University.

See more of this Session: Oxycombustion of Coal - Need, Opportunities, and Challenges II
See more of this Group/Topical: Topical E: High Temperature Environmentally Sustainable Energy Processes (sessions joint with the Environmental Division)

## Nitric Oxide Emissions during Coal and Coal/Biomass Combustion under Air-Fired and Oxy-Fuel conditions

*Melissa L. Holtmeyer, Scott A. Skeen, Benjamin M. Kumfer and Richard L. Axelbaum**
*Department of Energy, Environmental, and Chemical Engineering*
*Consortium for Clean Coal Utilization*
*Washington University in St. Louis*
*St. Louis, MO 63130*

### ABSTRACT

Co-firing coal with biomass is becoming an increasingly important approach to the development of renewable energy sources for electricity production, especially in regions that do not have ample sources of wind and solar power. Cofiring may also result in lower emissions of pollutants, including nitrogen oxides (NOx). In this study of the co-firing of pulverized coal and sawdust under air-fired and oxy-fuel conditions, we investigate the effects of biomass particle size and nitrogen content on nitric oxide (NO) formation. Experiments were performed in a 30 kWth laboratory-scale coal combustor. Measurements of NO concentrations in the exhaust stream revealed that the NO formation is dependent upon the sawdust particle size. When compared to pure coal combustion, co-firing of up to 40% (wt.) sawdust that contained large particles (> 50 mesh) did not result in lower NO emissions, despite the much lower nitrogen content associated with the sawdust. The conversion of fuel-bound nitrogen to NO was found to increase as the percentage of sawdust co-fired also increased. However, reductions in NO were obtained when the sawdust particle size was reduced. Numerical simulation of the experiments show that for larger particles, the particles penetrate the flame and devolatilization is delayed, resulting in the release of fuel-bound nitrogen in a region that is more conducive to NO formation.

### INTRODUCTION

An important concern facing regions that are dependent upon coal for electricity is the adoption of renewable energy portfolio standards. Recognizing that many of these regions do not possess solar and wind resources sufficient to meet renewable energy targets, biomass co-firing may play an important role in future power generation. Cofiring may also result in reduced emissions of criteria pollutant, such as NOx and $SO_2$, from coal-fired power plants. When introducing biomass into a coal-fired boiler, the combustion characteristics and emissions can change depending not only on the type of biomass but also on fuel processing, such as size reduction. In this work, the dependence of NO emissions on biomass cofiring ratio, fuel nitrogen content and particle size is investigated experimentally and numerically. Numerical simulation is employed to provide a detailed analysis of the combustion zone and to better understand these effects. Since new or retrofit coal-fired plants may employ oxy-combustion technology as a means to carbon capture, biomass cofiring under oxy-combustion conditions is also investigated experimentally in this study. The NO emissions measured under oxy-combustion and air-fired combustion conditions are compared.

There is evidence suggesting that reductions in NOx emissions when co-firing may not scale with reductions in fuel-bound nitrogen [1]. The effect of biomass co-firing on NOx emissions is complicated and dependent upon many factors, including fuel particle size and composition, nitrogen content, feeding configuration, flame stability, standoff distance, and burner hydrodynamics, both flow field and particle interactions [2]. The relationship between particle size and NOx emissions has been investigated for coal combustion. Devolatilization and the release of nitrogen are dependent on

*Corresponding author email: axelbaum@wustl.edu, Tel: (314) 935-7560, Fax: (314) 935-7211

temperature, particle heating rate, residence time, and particle size [3]. Smaller particles in coal combustion systems have higher heating rates, which can increase devolatilization rates and decrease time for nitrogen release. Midkiff et al. [4] suggested that if tar generation within the particle is independent of particle size, then the amount of nitrogen-bearing tar that escapes to the gas is greater for smaller particles. Thus, the release of greater amounts of nitrogen into the reducing environment of the primary combustion zone (PCZ) can augment NOx emission reductions. Bi et al. [5] showed that the level of NOx reduction significantly increases with decreasing coal particle size when coal is used as a reburn fuel in staged combustion. These NOx reductions in coal combustion due to decreased particle size can be linked to alterations in the devolatilization zone within the PCZ.

While the effects of particle size on NOx formation in pulverized coal (PC) combustion has been well-studied, particle size effects during co-firing where there exists two separate fuels with different characteristics, is a nascent area of study. Casaca et al. [6] showed that when biomass is used as a reburn fuel, there is an optimal particle size range for peak NOx reduction. The authors hypothesized that smaller than optimal particles completed devolatilization too early, resulting in an insufficient reburn zone, while larger than optimal sized particles continued to devolatilize downstream of the reburn zone exposing the volatile-nitrogen to oxygen. Lu et al. [7] showed through entrained-flow and single-particle reactor experiments that volatile yields decrease with increasing particle size. In the experiments described herein, biomass particle size is varied by sieving and co-fired with PC in order to directly observe the impact on NOx.

## EXPERIMENTAL

Experiments were conducted in a cylindrical, horizontally-fired 30 kWth combustor designed with a 14 cm ID combustion section, 78 cm in length, followed by a 37 cm ID burnout section, 120 cm in length [8]. The secondary oxidizer (SO) flow was introduced tangentially, to create a swirling flow to assist in flame stabilization. Pulverized coal and wood sawdust were fed to the combustor using separate volumetric screw feeders (K-Tron and Schenk AccuRate) and were entrained into the primary oxidizer (PO) stream using an eductor. The two feeder outlets and the eductor were contained within an enclosure maintained at atmospheric pressure, and the air flowing into the fuel inlet of the eductor was controlled and measured. Under oxy-coal conditions flue gases were not recycled; rather, industrial grade $CO_2$ was used. The refractory lined combustion chamber was sealed and maintained at a slight negative pressure (*ca.* 10 Pa). No buoyancy-induced asymmetries in the flow field were observed.

Gas samples were extracted from the exhaust at a rate of 200 cc/min using an uncooled ceramic 2 cm I.D. probe and were filtered, dried and diluted by a dilution probe system (Thermo Electron, Inc.). The concentrations of $O_2$, CO and NO were measured using a continuous emissions monitor (CEM) (Thermo Electron, Inc.). Nitrogen oxide concentrations were measured using a chemiluminescence analyzer (Thermo Electron, model 42i) and CO concentrations were measured using gas filter correlation (Thermo Electron, model 48i).

In order to compare NO measurements under air-fired and oxy-fuel conditions the measured concentrations must be normalized. In this work, the normalization is performed using the following equation.

$$NO\left[\frac{ng}{J}\right] = \frac{NO[ppm] \times \text{Molar Flow of Dry Exhaust } [kmol/s] \times MW_{NO}[kg/kmol] \times 10^6}{\text{Thermal Input } [W]}$$

Subbituminous Powder River Basin (PRB) coal and two different sawdust samples obtained from a sawmill (Sawdust A) and cabinetry manufacturer (Sawdust B) were utilized as fuels in this study. Sieve, proximate, and ultimate analyses were performed for both the coal and wood dust and the results are provided in Tables 1 and 2. Some variation in the coal particle size distribution was observed over the course of experiments as indicated in the table. As obtained from the sawmill, Bulk Sawdust A (< 18 mesh) contained 21 wt.% moisture, which led to feeding difficulties; however, after reducing the moisture content to 16 wt.% by exposing the sawdust to laboratory room air for a period of time, successful feeding was achieved. With further drying and time between experiments, Sawdust A (<50 mesh) contained 12 wt.% moisture. The moisture content of Sawdust B was approx. 7 wt.% as received. Measurements of the moisture content of both the coal and wood dust were taken daily prior to each set of experiments using a Hi-Res TGA (TA Instruments Inc.).

**Table 1. Fuel Properties: Ultimate & Proximate Analyses**

| | PRB Coal | Sawdust A | Sawdust B |
|---|---|---|---|
| HHV MJ/kg | 29.7 | 20.2 | 19.2 |
| **Proximate Analysis wt. % (dry)** | | | |
| Ash | 7.5 | 0.6 | 0.2 |
| Volatile Matter | 43.4 | 84.5 | 83.7 |
| Fixed Carbon | 49.1 | 14.9 | 16.1 |
| **Ultimate Analysis wt. % (dry)** | | | |
| Carbon | 69.51 | 49.28 | 48.80 |
| Hydrogen | 4.61 | 5.79 | 5.98 |
| Oxygen | 17.02 | 44.14 | 44.87 |
| Nitrogen | 0.97 | 0.15 | 0.01 |
| Sulfur | 0.40 | 0.05 | 0.11 |

**Table 2. Fuel Properties: Particle Size**

| | PRB Coal | Sawdust A | Sawdust B |
|---|---|---|---|
| **Sieve Analysis wt. % retained** | | | |
| 18 Mesh | 0 | 0 | 0 |
| 30 Mesh | <0.1 | 2.3 | -- |
| 40 Mesh | -- | -- | 21.8 |
| 50 Mesh | <0.1 | 36.3 | 43.8 |
| 70 Mesh | -- | -- | 63.7 |
| 100 Mesh | 1.9-3.2 | 81.3 | 79.8 |
| 200 Mesh | 18-29 | 94.7 | 93.5 |

To study the effect of biomass particle size on NO formation when co-firing with coal, two samples for each wood dust were prepared with different size distributions. For both sawdust samples a batch was prepared by removing only particles larger than 18 mesh (1 mm). These are referred to as $A_{bulk}$ and $B_{bulk}$. Batches of smaller particle sizes were prepared by sieving samples A and B through 50 mesh (0.3 mm) and 70 mesh (0.212 mm) screens, respectively. These are referred to as $A_{<50mesh}$ and $B_{<70mesh}$. Noting also that A contains 0.16 wt.% nitrogen and B contains 0.01 wt.% nitrogen (see Table 1), the effect of biomass nitrogen content on NO emissions during co-firing can be evaluated by co-firing both sawdust samples under similar conditions. In all experiments, conditions were set such that the thermal input was 30 kW and the exhaust gas oxygen concentration was maintained at 3 vol.% $O_2$. The effect of overfire air (i.e. oxidizer staging) was not investigated.

## NUMERICAL METHODS

The CFD program used was ANSYS FLUENT version 12.0. This program allows for user-defined modifications to the routine sub-models to include both coal and biomass combustion tracking. Transport equations were solved for the continuous (gaseous) phase in the Eulerian frame of reference, while simulating the discrete phase in the Lagrangian frame of reference. For this study, all discrete phase particles were assumed spherical. Coal and biomass reactions were modeled using the species transport sub-model. The finite-rate/eddy dissipation model, wherein both the Arrhenius and eddy-dissipation reaction rates are calculated, was applied for turbulent chemical reactions. The SST k-ω turbulence model was implemented, because of its abilities to capture swirling flow behaviors. The P1 radiation model was implemented in both conditions. A two-dimensional axisymmetric computational domain was used. Details about the individual sub-models are listed in Holtmeyer et al., [9].

## RESULTS

*Experimental*

The effect of co-firing sawdust on NO emissions is shown in Figures 1a and 1b. Air-fired conditions at 20 kWth are designated by the open triangles, air-fired conditions at 30 kWth are designated by the filled triangles, and oxy-fuel conditions at 30 kWth with a 30/70 volumetric mixture of $O_2$ and $CO_2$ in the PO and SO streams and a PO flow rate of 4.2 $m^3/hr$ are designated by the filled circles. Under air-fired conditions at 20 kWth the flame stand-off distance increased as the percentage of sawdust co-fired was increased. The lower SO flow associated with the 20 kWth flames results in a weak swirl and less recirculation Also, while the sawdust is more volatile than PRB coal, the larger particle size leads to less volatiles in the near burner zone. The resulting increased flame standoff distance, which could be visibly observed, is believed to result in entrainment of SO into the near-burner zone, thereby increasing the oxygen concentration in the primary devolatilization zone, leading to increased NO formation.

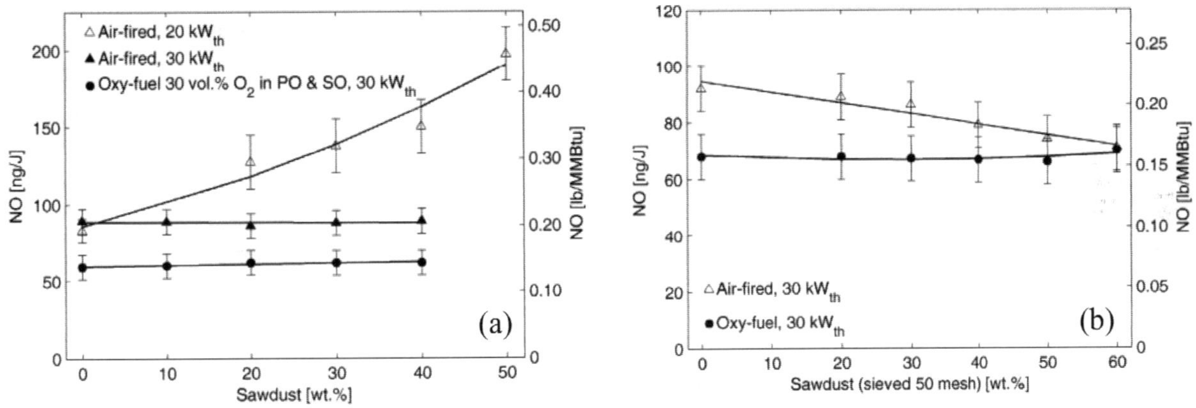


**Figure 1. (a) NO emissions vs. mass percentage of Sawdust $A_{bulk}$ co-fired with coal, under air-fired and oxy-combustion conditions. (b) NO emissions when cofiring with reduced sawdust particle size (sawdust $A_{<50mesh}$).**

When co-firing sawdust at a thermal input of 30 kW under air-fired conditions, the flame remained attached to the burner. However, as can be seen in Figure 1a, the NO emissions were not reduced even though the nitrogen content of the sawdust was 0.15 wt.% compared with 0.97 wt.% coal. Given the much lower nitrogen content in the sawdust, one might expect that the NO concentration in the exhaust gas should be less. The fact that it is relatively constant indicates that other aspects of the sawdust are contributing to NO formation.

Under oxy-fuel conditions when co-firing sawdust at 30 kWth the flame also remained attached. As with the air-fired 30 kWth flames, the amount of NO produced did not change when co-firing up to 40% sawdust. Moreover, these results show a 20% reduction in NO emissions under oxy-fuel conditions. Pershing and Wendt [11] found that *ca.* 80% of NO formation is due to fuel-bound nitrogen, and 20% is produced via the thermal mechanism. Thus, by eliminating NO with oxy-fuel the 20% reduction is consistent with eliminating the thermal NO pathway.

To understand the effect of biomass particle size on NO emissions, particles larger than approx. 300 microns were removed from the sawdust using a 50 mesh sieve prior to co-firing. The results of co-firing this sieved sawdust under air-fired and oxy-fuel conditions at 30 kWth are shown in Figure 1b. NO emissions decreased by approx. 20% when co-firing 60 wt.% of sieved sawdust under air-fired conditions. Under oxy-fuel conditions NO emissions were unchanged. This suggests that conclusions regarding co-firing in air may not be directly applied to oxy-combustion.

Since thermal NO is negligible under oxy-fuel conditions, co-firing should have reduced the amount of NO based on the fuel-bound nitrogen. As seen in Figure 2, the conversion of nitrogen to NO increased with the percent co-fire. Aho and coworkers [12, 13] indicated that fuel-N oxidation is dependent on the oxygen-to-nitrogen ratio (O/N) in the fuel. This could at least partially explain the observed increase in conversion with co-firing, since the O/N ratio of the sawdust is much larger than that for coal. Moreover, the ratio of $NH_3$/HCN released from biomass devolatilization is generally higher than that of coal, which may lead to increased conversion [14-15]. As seen in Figure 2, at higher co-firing ratios, the conversion to NO was greater for $A_{bulk}$ than for $A_{<50mesh}$, indicating that sawdust particle size can affect the overall N conversion under these conditions.

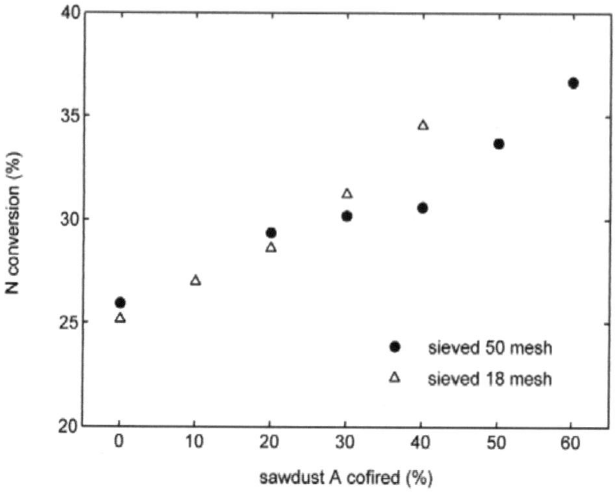


**Figure 2. Conversion of fuel-bound nitrogen to NO as a function of the percentage of sawdust co-fired. Sieved 18 mesh ($A_{bulk}$) and Sieved 50 mesh ($A_{<50mesh}$) are compared.**

**Table 3. Summary of NO emissions with variable particle size and fuel-N content**

| Fuel | N Content (wt.%) | NO (ng/J) |
|---|---|---|
| Coal only | 0.97 | 92 |
| **50 wt.% Cofiring** | **N Content of the Sawdust (wt.%)** | **NO (ng/J)** |
| Sawdust $A_{bulk}$ | 0.15 | 89 |
| Sawdust $B_{bulk}$ | 0.01 | 81 |
| Sawdust $A_{<50mesh}$ | 0.15 | 73 |
| Sawdust $B_{<70mesh}$ | 0.01 | 71 |

As shown in Table 3, co-firing $B_{bulk}$ at a co-firing ratio of 50%, produced 12% less NO than when co-firing with $A_{bulk}$. $B_{bulk}$ contains an order of magnitude less fuel bound nitrogen than $A_{bulk}$ (c.f. Table 1) and since the two bulk sawdust samples have similar particle size, the difference in NO emissions is attributed to the lower fuel-N content of Sawdust B. To again evaluate the effects of particle size, all particles greater than 70 mesh (0.212 mm) were removed from Sawdust B ($B_{<70mesh}$) by sieving. The NO measured in the exhaust was reduced by 23% at a co-firing ratio of 50% compared to 12% for $B_{bulk}$. As with Sawdust A, the larger particles yielded more NO than the smaller particles. From these results, reduced fuel-bound nitrogen coupled with sieved sawdust produces the lowest NO emissions. To further investigate this relationship between NO emissions and particle size, results from numerical simulation are presented and discussed.

*Numerical Simulation*

The temperature distributions obtained from the simulation are shown in Figure 3. Preliminary temperature measurements indicate that the location of peak temperature for the coal-only case compares well with the predicted location of peak temperature.

When co-firing, a high moisture content of the biomass has been shown to lead to a reduction in peak temperature. For this study, the sawdust had a moisture content similar to that of the coal, and the numerical results shown in Figure 3 indicate that peak temperatures are similar and independent of

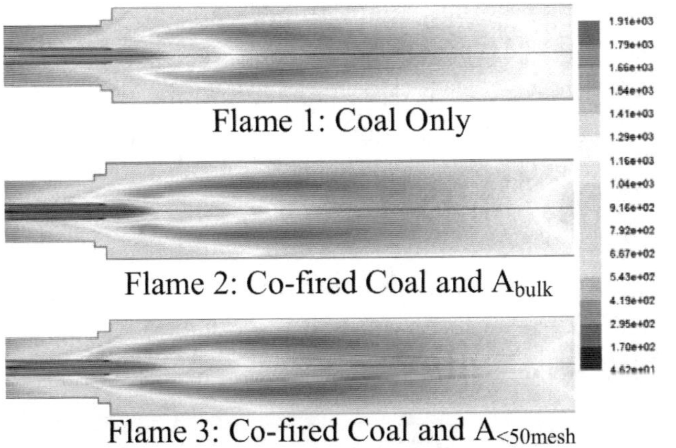


**Figure 3. Predicted temperature distribution for three flames: (1) coal only, (2) 50% co-firing $A_{bulk}$, (3) 50% co-firing $A_{<50mesh}$.**

particle size. However, the location of peak temperature differs. The larger size of the biomass contributes to changes in the overall flame structure, which not only affects the downstream peak temperature region of the flame, but also the near burner region. Similar changes were observed by Damstedt [17].

Figure 4a and 4b show particle tracks colored by particle diameter, and volatile concentration between coal and co-firing cases. Small particles, due to their relatively low momentum, stay entrained into the surrounding swirling flow after exiting the burner and are eventually transported away from the centerline. On the other hand, larger particles remain in the region of jet flow near the axis. Larger particles, due to their larger mass, take more time to reach the critical temperature at which volatiles are released. Robinson and coworkers [16] noted that biomass devolatilization times can be long due to intra-particle temperature gradients in the relatively large biomass particles. A longer high temperature zone, as seen in Figure 3, can be attributed to the slower heating of the large particles and the delayed release of associated volatiles. When the larger biomass particles are sieved out, the flame structure (Figure 3, Flame 3) is similar to that of Flame 1 as are the particle tracks.

The delay in volatile release from the larger biomass particles can be observed in Figure 4b. The accumulation of volatiles originating from the smaller biomass particles occurs in the same vicinity as the coal volatiles, however, volatiles originating from larger biomass particles are not released until further downstream. In Flame 1, coal particles have a much smaller range of particle sizes, so volatile release occurs approximately in the same region for all particles, resulting in a fuel-rich region near the exit of the primary oxidizer. The large range of particle sizes associated with the bulk co-firing flames results in a greater distance where volatiles are released. It is expected that the presence of large biomass particles and the associated delay in the release of volatiles is a major factor contributing to NO formation. Figure 5a and 5b illustrate the changes in NO formation that occur when co-firing.

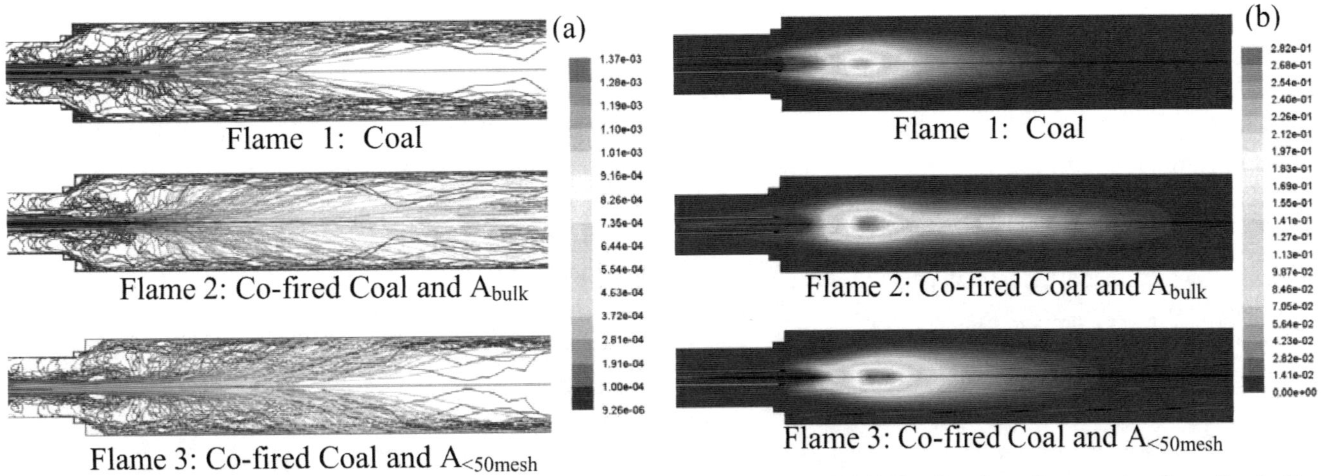


**Figure 4. (a) Predicted particle tracks colored by particle diameter and (b) Distribution of concentration of volatiles from parent fuel both for three flames: (1) coal only, (1) 50% co-firing of $A_{bulk}$ and (2) 50% co-firing of $A_{<50mesh}$**

The rate of fuel-NO formation normalized by the amount of fuel-bound nitrogen also occurs over a longer region in the co-fired flames (Figure 5a). The larger biomass particles associated with $A_{bulk}$ are shown to have a negligible effect on NO emissions. One would expect more NO formation to occur along the centerline downstream of the near burner region. Instead, volatiles are released further downstream in a region with less mixing and less exposure to higher oxygen concentration. Therefore, the changes in flame structure that occur during co-firing, as a consequence of the larger sawdust particles, could lead to higher fuel-NO emissions. During NO formation, volatile-N and char-N, in general, initially convert to intermediate species such as $NH_3$ or HCN. Focusing on the near burner region, the volatiles, once released from the particle, are entrained into the flow field. The internal recirculation zone, which entrains hot products, could bring volatiles in contact with inlet gases. Therefore, a greater conversion of intermediate species (HCN, $NH_3$) to NO due to the delayed release of volatiles in zones of higher oxygen concentration could result.

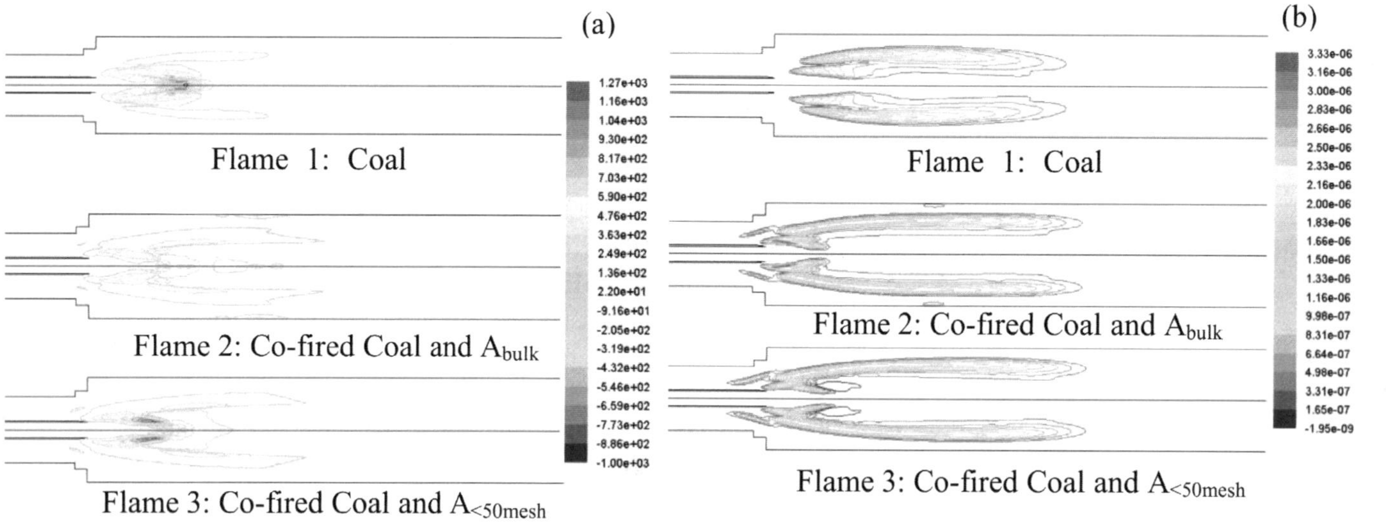


**Figure 5. (a) Contours of the rate of fuel-NO formation normalized by the amount of fuel-bound nitrogen, and (b) Contours of the rate of thermal-NO formation both for three flames: (1) coal only, (2) 50% co-firing of ($A_{bulk}$) and (3) 50% co-firing ($A_{<50mesh}$).**

Figure 5b shows the contours of the rate of thermal NOx formation. The co-firing flames have elongated regions of thermal NO formation, which correspond to the elongated high temperature regions seen in Figure 3. Nitrogen has a greater potential for conversion to NO in Flames 2 and 3, since the nitrogen is exposed to higher temperatures for longer periods of time.

Table 4 shows that the trends predicted by the model are consistent with the experiments on a ppm basis. During co-firing with $A_{bulk}$ (Flame 2), the flowrates were increased due to the higher fuel flowrates. Thus, while the concentrations of NO varied between Flames 1 and 2, there is little difference in the NO on a ng/J basis.

**Table 4. NO concentrations in exhaust from experiment and numerical simulations**

| Flame | Predicted (ppm) | Experiment (ppm) | Experiment (ng/J) |
|---|---|---|---|
| (1) | 296 | 285 | 92 |
| (2) | 282 | 242 | 89 |
| (3) | 174 | 227 | 73 |

## CONCLUSION

In this study of co-firing sawdust with coal, NO formation was found to be dependent on biomass particle size and nitrogen content. Experimental data showed that NO emissions independent of sawdust percentage, when co-firing bulk sawdust, remained relatively constant under air-fired and oxy-combustion conditions even though the fuel-bound nitrogen was significantly reduced in some cases. This indicated that other aspects of the sawdust are contributing to the NO formation. The effects of biomass particle size on NO emissions were investigated and it was found that sieving the sawdust reduced NO emissions. The lowest NO emissions were observed when cofiring with sieved sawdust having an order of magnitude lower fuel-bound nitrogen than coal.

Numerical simulation results showed that particle size effects and moisture content of the biomass are affecting the fuel-NO and thermal-NO pathways. When the combustion of small and large sawdust particles is compared, it was found that the release of volatiles associated with the large particles is delayed. This delay can potentially cause the intermediate nitrogen-containing species to be released into areas of higher oxygen in the near burner region. The model also shows that larger sawdust particles, due to their high momentum, penetrate through the fuel-rich near burner zone creating a longer high temperature region. When co-firing sawdust fuels with similar moisture content to the coal, the longer high temperature region coupled with similar peak temperatures, may promote the formation of thermal NO.

Although biomass may be low in nitrogen as compared to coal, significant reductions in NO emissions may not be realized, even for high co-firing ratios, unless strategies to reduce particle size or to develop burners specifically for low-NOx co-firing are adopted.

## ACKNOWLEDGEMENTS

This was funded by the Consortium for Clean Coal Utilization at Washington University supported by Peabody Coal, Arch Coal, and Ameren UE. MLH was supported under the U.S. EPA STAR Fellowship.

## REFERENCES

[1] J. Brouwer, W. D. Owens, N.S. Harding, M.P. Heap, D.W. Pershing, Co-firing waste biofuels and coal for emissions reduction. *Biomass Conference of the Americas: Energy, Environment, Agriculture and Industry, 2nd, Portland, Oreg., Aug. 21-24, 1995* 1995, 390-399.

[2] L. S. Pedersen, D. J. Morgan, W. L. van de Kamp, J. Christensen, P. Jespersen, K. Dam-Johansen, *Energy & Fuels* 11 (2) (1997) 439-446.

[3] S. C. Hill, L. Douglas Smoot, *Prog. Energy Combust. Sci.* 26 (4-6) (2000) 417-458.

[4] K. C. Midkiff, R. A. Altenkirch, K. Li, Proc. Combust. Instit. (U.S.) Eastern Section. Fall Technical Meeting, 1985.

[5] M. Bi, Y. Jia, *J. Therm. Sci.* 17 (3) (2008) 267-274.

[6] C. Casaca, M. Costa, *Proc. Combust. Instit.* 32 (2) (2009) 2641-2648.

[7] H. Lu, E. Ip, J. Scott, P. Foster, M. Vickers, L. L. Baxter, *Fuel In Press*, Uncorrected Proof (2008).

[8] S. Skeen, B. Kumfer, R.L. Axelbaum, Proc. 34$^{th}$ International Technical Conference on Clean Coal & Fuel Systems. Clearwater, FL. June 2009.

[9] M. Holtmeyer, S. Skeen, B. Kumfer, R.L. Axelbaum, Proc. 35$^{th}$ Internaional Technical Conference on Clean Coal & Fuel Systems Clearwater, FL. June 2010.

[10] D. W. Pershing, J. O. L. Wendt, Proc. *Combust. Instit.* 16 (1977) 389-99.

[11] M. J. Aho, J. P. Hämäläinen, J. L. Tummavuori, *Combust. Flame* 95 (1-2) (1993) 22-30.

[12] J. P. Hämäläinen, M. J. Aho, *Fuel* 74 (12) (1995) 1922-1924.
[13] K. R. G. Hein, J. M. Bemtgen, *Fuel Proc. Technol.* 54 (1-3) (1998) 159-169.
[14] E. E. Hughes, D. A. Tillman, *Fuel Proc. Technol.* 54 (1-3) (1998) 127-142.
[15] P. Glarborg, A. D. Jensen, J. E. Johnsson, *Prog. Energy Combust. Sci.* 29 (2) (2003) 89-113.
[16] J. Werther, M. Saenger, E.U. Hartge, T. Ogada, Z. Siagi, Combustion of agricultural residues. *Prog. in Energy and Combust. Sci.* 2000, 26, (1), 1-27.
[17] B. Damstedt. *Structure and Nitrogen Chemistry in Coal, Biomass and Co-firing Low $No_x$ Flames.* 2007. Doctoral Dissertation submitted to Brigham Young University

# (487g) Sulfur Release and Sulfur Capture in N2/O2 and CO2/O2 Fluidized Bed Combustion

Liyong Wang and Eric Eddings, Chemical Engineering, The University of Utah, Salt Lake City, UT

Fluidized bed combustors are very efficient systems for direct sulfur emission capture by limestone and related materials, and given the current interest in oxycoal combustion as an option for CO2 recovery, it is worthwhile to study sulfur release & sulfur capture in both O2/N2 and O2/CO2 conditions in fluidized bed systems. Sulfur release with and without limestone was investigated in N2/O2 and CO2/O2 environments for the following conditions: temperature range (600℃ - 870℃), O2 concentration range (10% - 30%), and a wide range of Ca/S ratios. In addition, a single coal particle combustion model was developed to help understand the mechanisms of sulfur release, sulfur capture and carbon release during coal combustion process. This paper will present results of both the detailed experimental study, as well as the companion single-particle modeling study, to provide explanations for the observed behavior of sulfur during oxy-coal combustion.

See more of this Session: Oxycombustion of Coal - Need, Opportunities, and Challenges II
See more of this Group/Topical: Topical E: High Temperature Environmentally Sustainable Energy Processes (sessions joint with the Environmental Division)

# (552a) Large Eddy Simulation of Axial Pulverized Coal Flames in O2/CO2 Environments

**Julien Pedel**[1], Jeremy Thornock[2] and Philip J. Smith[2], (1)Department of Chemical Engineering, University of Utah, Salt Lake City, UT, (2)University of Utah, Salt Lake City, UT

Oxy-coal combustion, in which an O2/CO2 mixture replaces air, is one of the few possible capture technologies to enable CO2 sequestration for existing coal-fired boilers. Burning coal with relatively pure oxygen, together with recycled flue gases, can produce a highly concentrated (up to 95% CO2) flue gas stream, which makes carbon sequestration more economical. One issue of interest in rapidly implementing a strategy to retrofit existing air-fired burners is to understand how replacing air by a O2/CO2 mixture affects the kinetics, aerodynamics and flame ignition in the near burner region. This study presents Large Eddy Simulations (LES) potential to predict oxy-coal flame characteristics and to be an important tool in the retrofitting process. In a turbulent coal flame, accurate predictions of the particle size and velocity distributions are critical for predicting the flame characteristics. Stokes number analyses suggest that LES has the capabilities to resolve all the turbulent length scales affecting coal particles in a coal flame, and could therefore lead to more accurate simulations. The Direct Quadrature Method of Moments was used to track the particle phase in an Eulerian framework. Simulations results show predictions of both the gaseous and solid phase dynamics and kinetics. The effect of partial pressure of oxygen on flame stand-off distance was investigated.

See more of this Session: Oxycombustion of Coal - Need, Opportunities, and Challenges III
See more of this Group/Topical: Topical E: High Temperature Environmentally Sustainable Energy Processes (sessions joint with the Environmental Division)

# Evaluation of gas radiation modeling in oxy-fired furnaces

*Stefan Hjärtstam*, Robert Johansson, Klas Andersson, Filip Johnsson*
*Department of Energy and Environment, Chalmers University of Technology, Sweden*

## Abstract

Oxy-fuel fired furnaces will exhibit combustion conditions different from air-firing and this will have important effects on the radiative properties of the gas. In contrast to nitrogen ($N_2$), carbon dioxide ($CO_2$) and water vapor ($H_2O$) are strong infrared emitters and the radiative activity of the gas in oxy-fuel combustion is increased compared to air-firing. In an oxy-fired furnace the pressure path-lengths are several times larger than in an air-fired furnace and the ratio of $H_2O$ to $CO_2$ can be significantly different. These differences between air and oxy-fuel combustion make the role of radiative heat transfer by gases in oxy-fired combustion different to air-firing, even in cases where the temperature distribution is similar to that of air-fired combustion. These aspects are essential for design of oxy-fired furnaces and need to be accounted for when temperature and heat transfer conditions are determined by modeling. In comprehensive combustion models, for example CFD models, it is common to neglect the spectral variations of the gases and to treat the spectrum by a single average, i.e. a gray approximation. Approximate models frequently applied in combustion modeling, such as the Weighted-Sum-of-Gray-Gases (WSGG) model, are not suitable for oxy-fired boilers since their parameters are fitted to pressure path-lengths, and ratios of $H_2O$ to $CO_2$, typical for air-fired conditions.

Different approximations used to account for gaseous radiation in CFD-simulations are investigated: temperature predictions by a gray model and a non-gray formulation of a WSGG model are compared in air- and oxy-fired conditions. The WSGG model used in this work is suitable for oxy-fuel conditions since it accounts for various ratios of $H_2O$ to $CO_2$ and the parameters are fitted to a broad range of pressure path-lengths. The modeled case is a propane flame in Chalmers 100 kW oxy-fuel rig. Two flames with similar temperature distribution are modeled and compared: an oxy-fuel case with 27 vol.% oxygen in the feed gas and an air-fired reference case. To support the CFD-results, calculations of the radiative source term are carried out for a domain between two infinite plates with similar temperature and concentration profiles as in the CFD-simulations. In these calculations, the gray model and the non-gray formulation of the WSGG model are compared with a Statistical-Narrow-Band model. In addition, the role of radiative heat transfer is investigated by including soot radiation in the CFD-simulations to give an understanding of the relative importance of gaseous radiation when soot particles are present.

It is shown that a use of a non-gray approach is motivated when modeling oxy-fuel combustion since the gray model fails in predicting the source term, which affects the predicted temperature field in CFD calculations. Furthermore, the inclusion of soot radiation is more crucial for the modeled flames than the use of a more rigorous description of the radiative properties of the gaseous components.

* Corresponding author. Tel + 46 31 772 1442, fax +46 31 722 3592, stefan.hjärtstam@chalmers.se

## Introduction

The power generation sector is one of the largest emitters of greenhouse gases, where carbon dioxide ($CO_2$) released from combustion is the most important one. Since a large fraction of power generation is in fossil fuelled power plants with abundant resources of these fuels, there is a great need to find ways to reduce $CO_2$ emissions from these plants. The only way to drastically cut emissions from fossil fuelled power plants is to develop and apply carbon capture and storage (CCS) technologies. Therefore, several CCS technologies are currently developed and oxy-fuel combustion is one of these technologies which is believed to have a great potential for commercialization [1].

In oxy-fuel combustion the fuel is burnt with pure oxygen ($O_2$) instead of air (as in conventional combustion). To control the combustion temperature, flue gas is recycled and reintroduced through the burner into the combustion zone. Hence, the combustion gases mostly consist of $CO_2$ and water vapor ($H_2O$), which makes an efficient capture of $CO_2$ possible. The replacement of nitrogen ($N_2$) with $CO_2$ and $H_2O$ affects several fundamental processes. The aerodynamics in the combustion chamber is different, the reaction chemistry in combustion is affected and the radiative heat transfer differs from that of air-firing. The latter, since $N_2$ is transparent to thermal radiation, but both $CO_2$ and $H_2O$ absorb and emit radiation at thermal wavelengths [2]. A proper heat transfer description of the oxy-fuel process is important for the design of future combustion units. This paper therefore focuses on the radiative heat transfer in modeling of oxy-fuel combustion.

The design of combustion devices is governed by heat transfer mechanisms. Radiative heat transfer is often dominant in combustion chambers and to predict this heat transfer the radiative transfer equation (RTE) needs to be solved. Several gases (mostly $CO_2$ and $H_2O$) within the combustion chamber have spectrally dependent properties consisting of millions of lines, and a complete integration over the spectrum is complicated. Modeling of combustion in industrial scale devices with calculations that take all spectral lines into consideration (line by line calculations (LBL)) is too complex. In simplified radiation models it is instead common to treat the spectrum by a gray approximation (an average of the whole spectrum) [3, 4]. In addition to LBL and gray models there are other approximate models that take the spectral properties of the gases into consideration. For instance, narrow-band and wide-band models estimate the emissivity or transmissivity of spectral bands of the physical spectrum. The drawback with these band models is that they require a correlated formulation of the RTE and are therefore computationally demanding. There are also absorption-coefficient models that treat a distribution function of the absorption coefficient for a spectral band or for the whole spectrum and these are computationally more efficient. The banded formulation of the Weighted-Sum-of-Gray-Gases (WSGG) model [5] is the most simplified formulation of the global absorption-coefficient models. Parameters for the WSGG model are typically valid for a broad temperature range but only for certain molar ratios of $H_2O$ and $CO_2$ [6-8]. Furthermore, the existing parameters of these models are often limited to specific pressure path-lengths. Recently, Johansson et al. [4] modified the WSGG model and new parameters were presented to enable the model to handle a span of molar ratios of $H_2O$ and $CO_2$ between 0.125-2. These parameters are also applicable to larger pressure path-lengths than previously presented parameters. The major reason for the new set of parameters was to make the model applicable to oxy-fuel combustion conditions in large scale boilers.

There are relatively few CFD based studies on radiation in oxy-fuel combustion availabe in the literature. Porter et al. [2] performed CFD calculations, with the P1 method to solve the radiative transfer equation, on two fictive methane-fired flames (one oxy-fuel and one air) using the standard Gray WSGG model and the Full-Spectrum k-distribution method (FSCK). In terms of temperature along the centerline of the furnace, the gray model and the FSCK method gave similar results for the oxy-fuel flame, but when comparing incident radiation the FSCK method predicted higher values than the gray approach. Porter et al. concluded that the reason for the similar temperature profiles is that the gray model underestimates the emission by hot gases because the absorption coefficient remains low, despite high levels of $CO_2$. They conclude that new total emissivity data with corresponding coefficients adopted for oxy-fuel conditions could lead to improvements, but any gray approximation is believed inferior to the FSCK method. The impact of soot on the radiation was also addressed in the study with the conclusion that the relative impact of soot in the calculations of temperature was lower in the oxy-fuel flame than in the air flame, due to a higher absorption coefficient of the gases in the oxy-fuel case. In a follow up paper by Porter et al. [9] the Discrete Ordinates method was used to solve the radiative transfer equation in CFD calculations for two air flames and one oxy-fuel flame. The standard gray WSGG model and the FSCK method used in the CFD modeling were compared with reference model in terms of a statistical narrow band model. The FSCK gave promising predictions of the radiative properties, but the gray model gave substantial errors in the calculation of the radiative source term. The authors concluded that it is essential to use a non-gray model for the radiative properties when modeling oxy-fuel combustion.

The above mentioned work give valuable contribution to the understanding of the radiative heat transfer phenomena in oxy-fuel combustion. Yet, there is a need to investigate how a WSGG model, developed for oxy-fuel conditions, behave when used in CFD calculations. This paper aims to investigate aspects concerning radiation modeling in oxy-fuel flames. First, the performance of the modified WSGG model by Johansson et al. [4], treated both as non-gray and gray, is investigated when used in a computational fluid dynamics (CFD) framework for an oxy-fuel environment. An additional objective in this work is to compare the importance of gas radiation representation with the impact of soot radiation. CFD calculations of both an oxy-fuel flame and an air flame in Chalmers 100 kW test unit [10] are carried out using both a non-gray formulation and a gray formulation of the modified WSGG model. Furthermore, CFD calculations with the non-gray formulation of the gas radiation model including gray soot radiation are carried out. The soot load is based on concentrations fitted by comparing the modeled radiation intensity with measured intensity profiles of the flames. To further evaluate the CFD-results, analytical calculations of the radiative source term are carried out for a domain between two infinite plates with similar temperature and concentration profiles as in the CFD-simulations, with the purpose to discuss the prediction accuracy of the source term by different gas radiation models. In these calculations, the gray and the non-gray WSGG models are compared against a reference model in terms of the Malkmus Statistical-Narrow-Band model [11].

## Theory

This part covers the treatment of radiation in the CFD analysis and the analytical calculations. In this study a gray formulation of the modified WSGG model (hereafter termed Gray WSGG model) [4], a banded formulation of the same WSGG model (termed Non-Gray

WSGG model) and the Malkmus Statistical Narrow-Band model (termed SNB model) [11] are used for this purpose. In an infinite plates calculation all three models are compared using the SNB model as a reference model, since it has shown good agreement with line by line (LBL) calculations [12]. Furthermore, the parameters of the WSGG model were obtained by fitting total emissivities to data predicted by the SNB model. Radiative heat transfer is included as a source term in the energy equation in CFD calculations. The source term is the difference between locally absorbed and emitted radiative energy.

***Solution of the radiative transfer equation***

The transport of radiative intensity through a medium is governed by the radiative transfer equation (RTE) shown in Eq. (1).

$$\frac{dI_\nu(\vec{r},\vec{s})}{ds} = -(\kappa_\nu + \sigma_{s,\nu})I_\nu(\vec{r},\vec{s}) + \kappa_\nu I_{\nu,b} + \frac{\sigma_{s,\nu}}{4\pi}\int_{\Omega=0}^{4\pi} I'_\nu(\vec{r},\vec{s}')\,\Phi(\vec{s}\cdot\vec{s}')d\Omega' \quad (1)$$

At a specific wavelength $\nu$, position $\vec{r}$ and in the direction $\vec{s}$ the intensity is weakened by absorption and scattering and strengthened by in-scattering and emission from the medium. An exact analytical solution of the RTE, which is an integro-differential equation, in a multi-dimensional case is complicated and hence only possible in simplified cases. Therefore, numerical methods which are able to approximate the solution of the RTE are necessary, for instance when modeling the radiation in a combustion chamber. There are different models available for this task and in this work a variant of the Discrete Ordinates Method (DOM) also known as the $S_N$-method has been applied [13, 14]. In the CFD calculations, performed in this work, care was taken in choosing an angular discretization detailed enough to generate a solution independent of the discretization. In the infinite plates calculations the Discrete Transfer Method has been applied and the $S_{12}$ discrete ordinates quadrature was used.

***Gaseous radiative properties***

As mentioned previously, an absorbing, emitting and scattering participating medium is characterized by its absorption coefficient and scattering coefficient. In this work only gaseous combustion is investigated and therefore scattering has been excluded in the calculations. Soot is included in some of the calculations, but the particles are assumed to be small and non-scattering.

The total gas emissivity correlation named the WSGG model, introduced by Hottel and Sarofim [15], is one of the models used in this study to describe the emissivity of the gas medium. The total emissivity, in Eq. (2), over a distance $s$, is with the WSGG model calculated as a summation over a small set of gray gases and one clear gas is [16]

$$\varepsilon = \sum_{i=0}^{I} a_{\varepsilon,i}(1 - e^{-\kappa_i S_m P(Y_{CO2}+Y_{H2O})}) \quad (2)$$

where, $a_{\varepsilon,i}$ is the corresponding weight factor (fraction of blackbody radiation) for the $i$ th fictitious gas and $\kappa_i$ is the absorption coefficient of the $i$ th gas. Values of $a_{\varepsilon,i}$ and $\kappa_i$ are usually dependent on the gas composition, and $a_{\varepsilon,i}$ also on temperature. When the WSGG model is used in CFD modeling it is common to make a gray gas approximation and calculate total radiative properties which are used for solving the spectrally integrated RTE. In order to estimate the absorption coefficient a characteristic length is needed. Usually, this is either the length of a computational cell or a characteristic length of the computational domain. In this work the domain-based approach has been used since the cell-based approach results in grid-dependent solutions [17].

With the Non-Gray WSGG model the approach is slightly different. Here, a RTE for each of the gray gases and the clear gas is solved. By multiplying the total blackbody radiation with the weight of each gas the emitted blackbody radiation for the respective RTE is obtained. When discretizing the RTE for gas $i$ using the non-correlated recurrence relation the RTE is given by

$$I_{i,n} = I_{i,n-1} \cdot e^{-\kappa_i \Delta s P(Y_{CO2}+Y_{H2O})} + a_{\varepsilon,i} I_{b,i,n-1/2} \cdot (1 - e^{-\kappa_i \Delta s P(Y_{CO2}+Y_{H2O})}) \quad (3)$$

In Eq. (3) $\Delta s$ is the length of the computational cell. The total intensity $I_n$ is calculated as the sum of the intensities of the $i$ gases. In the modified WSGG model by Johansson et al. [4] weights and absorption coefficients for four gray gases and one clear gas are presented. These coefficients were obtained by fitting to total emissivities calculated with the reference SNB model. The weights of the WSGG model are polynomial functions of temperature and the molar ratio of $H_2O$ to $CO_2$. The absorption coefficients are functions of the molar ratio solely. Compared with the standard WSGG model the advantage of the modified model is that it accounts for molar ratios of $H_2O$ to $CO_2$ ranging from 0.125 to 2 and pressure path-lengths up to 60 bar m.

In the SNB model the spectrum is divided into a large number of bands. Since the bands are narrow, the blackbody radiation can be treated as constant, which makes a statistical treatment possible. The reference model used in the infinite plates calculations is the Malkmus SNB model [11]. Band parameters and line half-width $\gamma$ correlations are from the work of Soufiani and Taine [18] and non-uniformity is accounted for with the Curtis-Godson approximation [19]. For each band in the SNB model the mean line-intensity to the typical line-spacing ratio $k_k$ is tabulated as well as the mean line-spacing $d_k$. For a narrow band $k$ the spectrally averaged transmissivity is calculated as

$$\bar{\tau}_{v_k} = exp\left[-\frac{2\gamma}{d_k}\left(\sqrt{1+\frac{Y_{specie}PSk_kd_k}{\gamma}}-1\right)\right] \quad (4)$$

In case the band includes both $H_2O$ and $CO_2$ the total transmittance of the band is the product of the two transmissivities. For each band an RTE is solved according to a correlated formulation and the total intensity $I_n$ is the sum of all band intensities

$$I_n = \sum_k \left(\Delta v_k \bar{I}_{v_k,0} \bar{\tau}_{v_k,0\to n} + \sum_{j=0}^{n-1} \Delta v_k \left(\bar{\tau}_{v_k,j+1\to n} - \bar{\tau}_{v_k,j\to n}\right) I_{bv_k,j+\frac{1}{2}}\right) \quad (5)$$

where $j$ describes the spatial discretization of the path. Index 0 refers to the background wall.

## Method

Two separate analyses of the radiative conditions in air and oxy-fired flames are applied. In the first part, CFD calculations are conducted to examine the influence of different gas radiation models on the temperature field of both air and oxy-fuel flames with comparison with experimental data. The relative impact of soot is also addressed in the CFD calculations. The second part includes analytical calculations of the radiative source term in a domain between to infinite plates.

***CFD analysis***

CFD modeling of propane-fired air and oxy-fuel flames are used to compare the Non-Gray WSGG model with the Gray WSGG model. To study the influence of soot on the two flames, additional CFD-calculations were performed including gray soot particles based on concentrations obtained by fitting modeled intensity profiles against measured profiles. These intensity profiles were presented in the work by Andersson et al. [20], who modeled and measured the intensity profiles of the Chalmers 100 kW flames (the same flames investigated in the present work) to study the radiation by soot in propane-fired oxy-fuel environments. By including soot in the CFD calculations the relative importance of both gas and soot radiation is investigated in the present work.

A commercial CFD-software (ANSYS FLUENT 12.1.4) was used for the calculations and both the Gray and the Non-Gray WSGG models were implemented as user defined functions (UDF). The CFD calculations were carried out in 3D with one quarter of the combustion chamber being modeled using periodic boundary conditions. The reason for the selection of one quarter of the furnace was to be able to include one of the four symmetrically placed cooling tubes in the simulation. The grid, constructed purely by hexahedral cells, is made dense in the near burner region where the actual flame is present and towards the outlet of the furnace the grid is coarser since no major reactions occur in this region (outside the flame). Care was taken in achieving grid independent solutions and this exercise resulted in the use of a structured grid, shown in Fig. 1, which consists of approximately 100 000 cells.

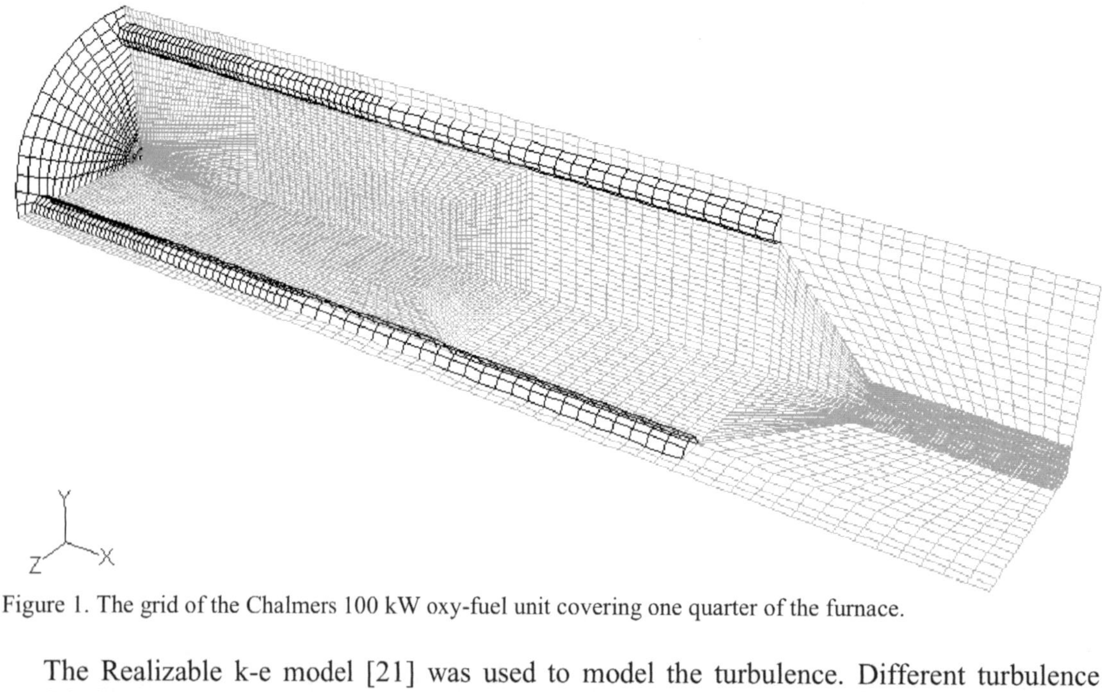


Figure 1. The grid of the Chalmers 100 kW oxy-fuel unit covering one quarter of the furnace.

The Realizable k-e model [21] was used to model the turbulence. Different turbulence models have been evaluated in a previous CFD study of the Chalmers 100 kW oxy-fuel unit [22], where the Realizable k-e model gave promising results. The chemistry model used was the global four reaction mechanism by Jones and Lindstedt [23]. Turbulence-chemistry interactions were modeled using the Eddy Dissipation Concept [24] which is able to account for slow reactions and equilibrium effects. Piecewise polynomial specific heat capacities are used for the participating species. For all simulations the segregated steady state solver is used. The pressure-velocity coupling is solved by the SIMPLE algorithm and transport scalars are discretized using a second order upwind scheme. The In-Situ Adaptive Tabulation (ISAT) algorithm [25] is used in combination with the EDC model to speed up the calculation time. Feed gas and fuel inlet boundary conditions are taken from experimental data. Wall temperatures are estimated from wall mounted thermocouples and measured gas temperature close to the wall.

### *Analytical calculations*

In addition to the CFD modeling, calculations of the radiative source term between two infinite plates are undertaken with the discrete transfer method. The purpose is to investigate the accuracies of the two WSGG models in describing the radiative source term in the energy equation. The conditions in the domain between the plates, in terms of temperature distribution and gas composition, are taken from the CFD-results. The SNB model is used as a reference model to validate the performance of both the Gray WSGG model and the Non-Gray WSGG model.

### *Validation data for the CFD simulations*

The experimental results modeled in this study were presented by Andersson et al. in 2005 [10]. Chalmers 100 kW combustion unit, shown in Fig. 2, is designed for oxy-fuel combustion with flue gas recycling. The cylindrical refractory lined furnace has an inner diameter and height of 0.8 and 2.4 meters, respectively. The unit is top-fired and cooled by four cooling tubes located symmetrically close to the furnace walls. In-flame measurements are made in seven measurement ports located along the reactor side.

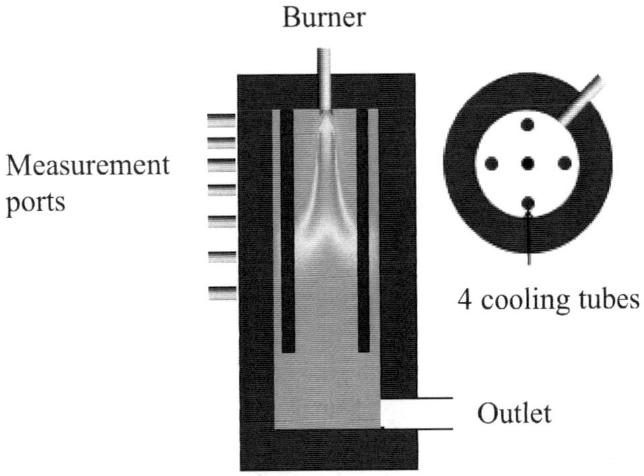


Figure 2. The combustion chamber of the Chalmers 100 kW oxy-fuel unit is the computational domain of the CFD calculations.

**Table 1.** Test conditions for the air case and the oxy-fuel case.

| **Heat input** | 80 kWth | **Fuel** | | $C_3H_8$ | **Stoich ratio** | 1.15 |
|---|---|---|---|---|---|---|
| **Test case** | **Combustion media** | **Temp feed gas [°C]** | **Feed gas composition (vol.%)** | | | **Theor. $CO_2$ conc. @ stack (vol.%)** |
| | | | $O_2$ | $N_2$ | $CO_2$ | |
| **Air** | Air | 25-30 | 21 | 79 | - | 12 |
| **OF 27** | $O_2/CO_2$ dry recycle | 25-30 | 27 | - | 73 | 96 |

Two different propane-fired flames, described in [10, 26], are modeled in the work presented here. One reference case, where air is used as oxidizer, and one oxy-fuel case (hereafter termed: OF 27 case). Both cases have a fuel input of ~80 kW and a stoichiometric ratio of 1.15. The OF 27 case has 27 vol.% $O_2$ in the feed gas and the remaining part consists mostly of $CO_2$, since dry recycling is applied. Test conditions for the two studied cases are given in Table 1. Both propane (98.3 mol % $C_3H_8$, 1.1 mol % $C_4H_{10}$) and oxygen (99.5 vol.% $O_2$) are fed from bottles.

Propane is fed through a gas lance located in the centre of the burner and leaves the burner via a small bluff body. The gas lance is surrounded by primary and secondary feed gas registers. Out of the total feed gas flow about 40 vol.% enters through the primary register and the remaining part enters via the secondary register. The swirl generating fin angle of the primary register was 45 degrees and for the more moderately swirling secondary register the fin angle was 15 degrees. Wall temperatures are measured by several thermocouples located along the side of the reactor ~20 mm from the inner furnace wall surface. For the OF 27 case the average wall temperatures, at an axial distance of 0-1400 mm from the burner, are ~585±50°C and for the air case ~595±70°C. The measured average gas temperature close to the wall (0- 1400 mm from the burner) were 732°C for both cases. In the numerical and analytical calculations, carried out in the present work, the wall temperatures were estimated to 627°C (0- 800 mm from the burner), 577°C (800-1800 mm from the burner) and 527°C (1800-2400 mm from the burner). For more experimental details see the work by Andersson et al. [26].

## Results

### *CFD analysis*

Figure 3 shows the modeled temperature along the centerline compared with measured data for the OF 27 case and the air case. For the OF 27 flame, shown in Fig. 3a, the Gray WSGG model yields the highest peak temperature but also the strongest temperature reduction along the centerline. The Gray WSGG model also predicts the highest peak temperatures in the air case, Fig. 3b, but the temperature reduction along the centerline is slower than what is shown by the experiments. In the OF 27 case the Non-Gray WSGG model estimates a slightly lower peak temperature, but a considerably longer zone with high temperature compared to the gray model. Hence, there is an improved agreement between measured temperature data and the modeling results with the Non-Gray WSGG model compared to when the gray approach is adopted. For the air flame the difference in the modeled temperature between the gray and the non-gray model is negligible. Inclusion of soot radiation in the Non-Gray WSGG model leads to a temperature reduction in the core of the flame where the soot is present. The soot, which contributes significantly to the radiation, as will be shown in the infinite plates calculations below, results in increased emission in the high temperature zone and thus a lower temperature. As a result, the modeled temperature along the centerline in the CFD calculations where soot is included follows the measurement data well for both the air and the OF 27 case.

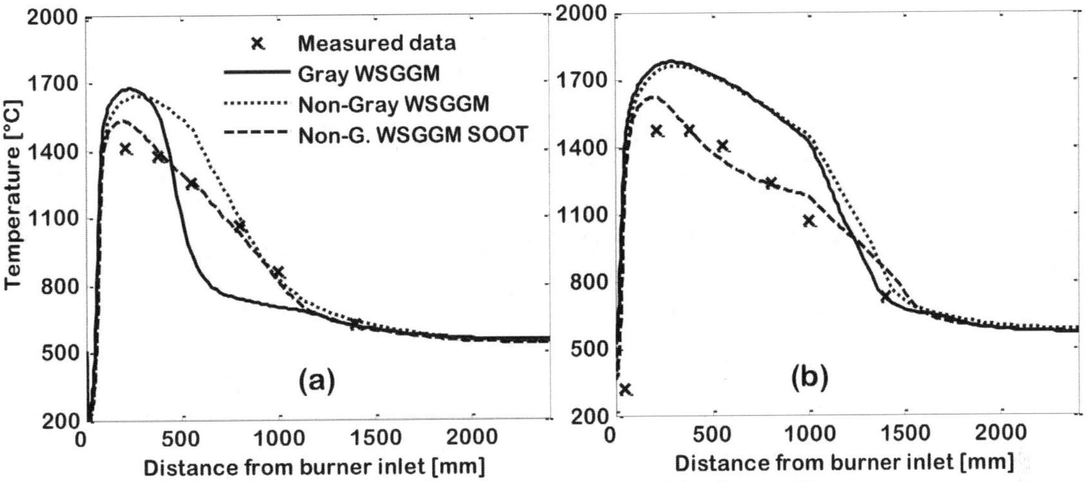


Figure 3. Temperature along the centerline in the OF 27 case (a) and the air case (b).

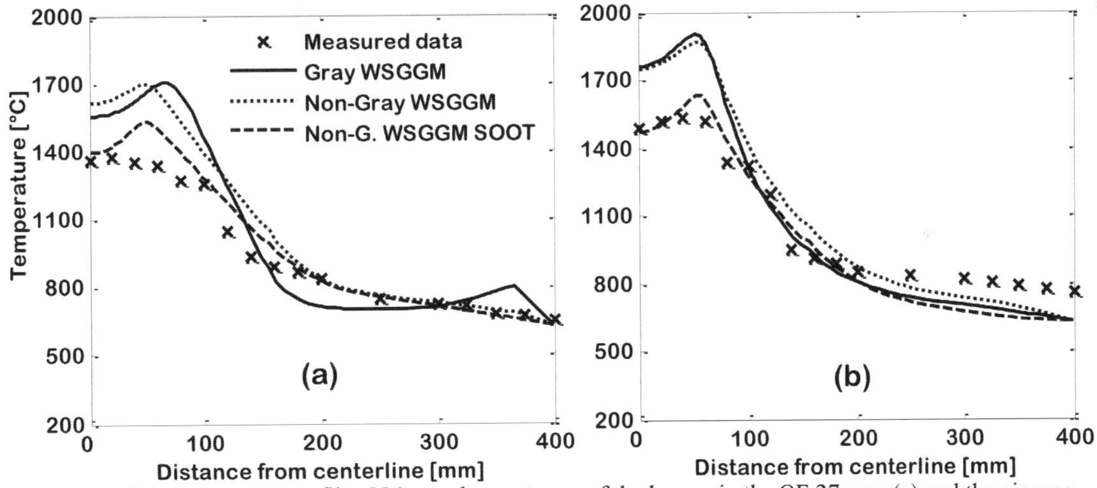


Figure 4. Radial temperature profiles 384 mm down-stream of the burner in the OF 27 case (a) and the air case (b).

Figure 4 gives modeled and measured temperatures for a radial profile located 384 mm down-stream of the burner. At this axial distance from the burner a visible flame is present and temperatures are therefore high. Similar to what was shown in the comparison with centerline data, the gray and the non-gray model predict almost identical temperature profiles for the air flame, given in Fig. 4b. However, in the OF 27 flame (Fig. 4a) the gray model predicts a slightly wider flame with a stronger temperature gradient than the Non-Gray WSGG model. Furthermore, the gray model predicts a temperature increase 250 mm from the centerline and out towards the wall, a prediction not seen in the measured data. The Non-Gray WSGG model is able to capture the principal shape of the measured temperature profiles of the OF 27 case except in the flame center (radial distance 0-100 mm) where the modeled temperatures are overestimated by approximately 250°C. Further improvements in the agreement with measurements is achieved with the Non-Gray WSGG model with soot included, although the model overestimates the peak temperatures by about 100°C in both flames.

The agreement with measured data is less satisfactory for the radial profiles located 215 mm down-stream of the burner, which is given for the OF 27 case in Fig. 5. Here, all models give a peak temperature significantly higher than the measured temperature. Furthermore, all models predict a temperature drop ~75 mm from the centerline. In the simulations, this temperature drop is due to the cold feed gas stream that penetrates this far into to combustion chamber. In addition to the radiation model there are several other possible reasons for the overestimated peak temperatures. Simple global reaction mechanisms, such as the mechanism used in this study, have a tendency of overestimating peak temperatures [27-29]. In addition, when performing measurements with a suction pyrometer, gas is sucked out from a relatively large volume, which results in that local temperature gradients are smoothened out. The fluctuations of the flame are also significant, which is something that the suction pyrometer cannot resolve. An insufficient description of the flow field by the turbulence model can also lead to deviations from measured data. Due to the above mentioned reasons the peak temperatures are overestimated and the temperature profile 215 mm from the burner deviates from measured data. To deepen the evaluation of the CFD-results, calculations of the radiative source term are carried out in the next section to investigate the prediction accuracy of the source term by the different gas radiation models.

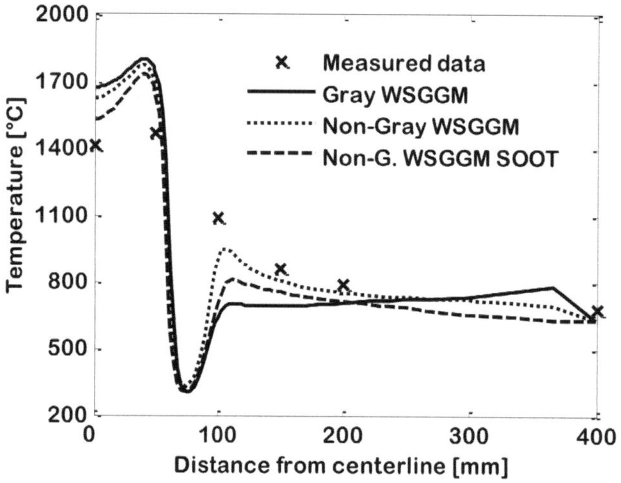


Figure 5. Radial temperature profiles 215 mm down-stream of the burner for the OF 27 case.

### *Analytical calculations*

The calculated radiative source terms for a domain between two infinite plates are shown in Fig. 6. The conditions in the domain are aimed to resemble the OF 27 case (Fig. 6a) and the air case (Fig. 6b), 384 mm down-stream of the burner. Temperatures and concentrations of $H_2O$ and $CO_2$ are set from the CFD calculations of the respective flames using the Non-Gray WSGG model. The length of the computational domain is the same as the diameter of the Chalmers 100 kW unit.

The source term of the OF 27 flame, presented in Fig. 6a, is negative in the center of the domain where the highest gas temperature is present. Close to the wall the source term is positive, since the colder gas absorbs radiation from the hotter central part of the flame. The

Non-Gray WSGG model agrees well with the reference model and should therefore give a good measure of the influence of gaseous radiation in the CFD simulations. The Gray WSGG model is, on the other hand, not able to capture the magnitude of the source term. Approximately 200 mm from the wall both the SNB and the Non-Gray WSGG model shows a positive peak in the source term. In this region the colder gas absorbs radiation from the center of the flame. The Gray WSGG model is not able to capture this absorption. The source terms of the air case, Fig. 6b, behave similar to the source terms of the OF 27 case. Again the Gray WSGG model fails to match the source term of the SNB model. Even though the temperature in the air case is higher than in the OF 27 case the magnitude of the source term is smaller, compared to the results presented in the OF 27 case where the gas emission and absorption is stronger. Despite the differences of the source term between the Gray and the Non-Gray WSGG model the two models give very similar temperature profiles in the CFD simulation of the air case and it is thus not always necessary to apply a non-gray formulation.

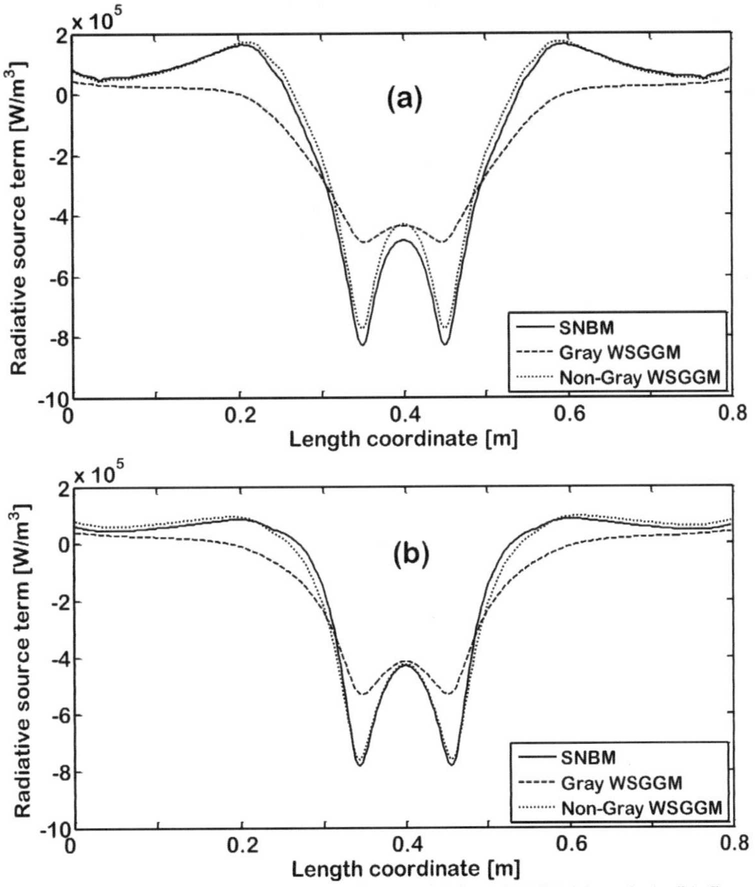


Figure 6. Predicted radiative source term for the Chalmers 100 kW OF 27 (a) and air (b) flames at an axial distance of 384 mm from the burner. The temperature and gas composition are taken from the CFD simulations performed with the Non-Gray WSGG model.

To study the influence of soot on the source term, additional infinite plate calculations were carried out for both the OF 27 and the air flame. Temperatures and gas compositions were taken from the CFD calculations with the Non-Gray WSGG model with soot included. Figure 7a yields similar results as the case presented in Fig. 6a, when the respective source terms of the Gray and the Non-Gray WSGG models are compared. In the OF 27 flame, Fig. 7a, the source term is strongly affected by the presence of soot as shown in the Non-Gray

WSGG model case including soot. The negative peak of the source term for the Non-Gray soot case is approximately 4 times the value the Non-Gray case without soot. This, together with the temperature differences noticed in the CFD simulations, shows that the inclusion of soot in the radiation model is of great importance for CFD simulations of sooty flames. From the analytical calculation of the source term as well as from the CFD analysis it is shown that the inclusion of soot radiation is more crucial for the result than the use of a more rigorous description of the radiative properties of the gaseous components.

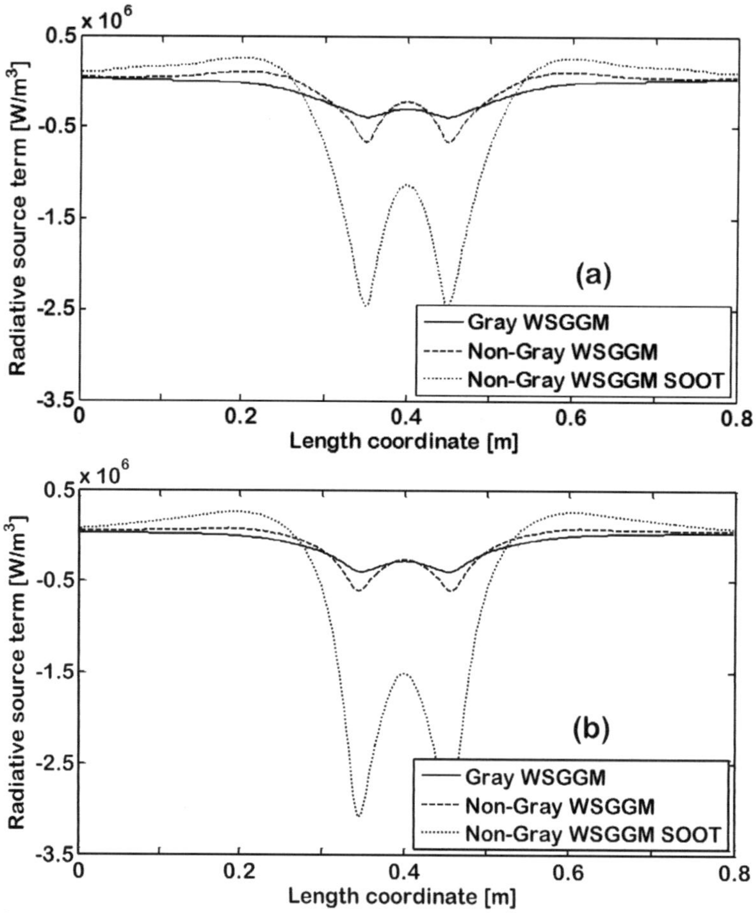


Figure 7. Radiative source term for the Chalmers 100 kW OF 27 (a) and air (b) flames at an axial distance of 384 mm from the burner. The temperature and gas composition are taken from the CFD simulations performed with the Non-Gray WSGG model with soot included.

## Conclusions

A WSGG model applicable to oxy-fuel conditions is evaluated by CFD and by analytical calculations. The WSGG model is treated both as gray and as non-gray. Modeled temperature profiles for both an oxy-fuel and an air flame are compared with measured temperature data. Furthermore, the impact of including soot in the Non-Gray model is investigated.

In the oxy-fuel flame the Non-Gray WSGGM predicts a larger high temperature zone with slightly lower peak temperatures compared to the Gray WSGG model. Compared with measured temperature data, both models overestimate peak temperatures but the non-gray model yields temperature profiles more in line with the measurements than the gray approach. In the air flame both models yield similar predictions of temperature. As in the oxy-fuel case, the peak temperatures are overestimated compared with measured data. When modeling the radiation in oxy-fuel combustion the use of a non-gray approach is motivated since the gray model fails in predicting the source term, which affects the predicted temperature field in CFD calculations.

The inclusion of soot in the Non-Gray WSGG model has a large impact in the CFD calculation, since the predicted flame temperatures are reduced by ~200°C. Soot radiation is more crucial for these flames than the use of a more rigorous description of the radiative properties of the gaseous components.

## References


1. Simbeck, D.R., McDonald, M. Existing power plant retrofit $CO_2$ control options analysis. $5_{th}$ International Conference on Greenhouse Gas Control Technologies, Cairns, Australia, 2000.
2. Porter, R., Pourkashanian, M., Williams, A., Smith, D. Modelling radiative heat transfer in oxycoal combustion. Proceedings of HT2009, ASME Summer Heat Transfer Conference, California, 2009.
3. Morvan, D., Porterie, B., Loraud, J.C., Larini, M. Numerical simulation of a methane/air radiating turbulent diffusion flame. International Journal of Numerical Methods for Heat & Fluid Flow 10 (2): 196-227, 2000.
4. Johansson, R., Leckner, B., Andersson, K., Johnsson, F. Account for variations in the $H_2O$ to $CO_2$ molar ratio when modelling gaseous radiative heat transfer with the weighted-sum-of-grey-gases model, submitted for publication.
5. Modest, M.F. The weighted-sum-of-gray-gases model for arbitrary solution methods in radiative transfer. Journal of Heat Transfer 113: 650-656, 1991.
6. Smith, T.F., Shen, Z.F., Friedman, J.N. Evaluation of coefficients for the weighted sum of gray gases model. Journal of Heat Transfer 104: 602-608, 1982.
7. Coppalle, A., Vervisch, P. The total emissivities of high-temperature flames. Combustion and Flame 49: 101-108, 1983.
8. Taylor, P.B., Foster, P.J. Some gray gas weighting coefficients for $CO_2$-$H_2O$-soot mixtures. International Journal of Heat and Mass Transfer 18: 1331-1332, 1975.
9. Porter, R., Liu, F., Pourkashanian, M., Williams, A., Smith, D. Evaluation of solution methods for radiative heat transfer in gaseous oxy-fuel combustion environments. Journal of Quantitative Spectroscopy & Radiative Transfer 111: 2084-2094, 2010.
10. Andersson, K., Johnsson, F. Flame and radiation characteristics of gas-fired $O_2$/$CO_2$ combustion. Fuel 86: 656-68, 2007.
11. Malkmus, W. Random Lorentz model with exponential-tailed $S^{-1}$ line-intensity distribution function. Journal of the Optical Society of America 57: 323-329, 1967.
12. Pierrot, L., Soufiani, A., Taine, J. Accuracy of narrow-band and global models for radiative transfer in $H_2O$, $CO_2$, and $H_2O$-$CO_2$ mixtures at high temperature. Journal of Quantitative Spectroscopy and Radiative Transfer 62: 523-548, 1999.
13. Chui, E.H., Raithby, G.D. Computation of Radiant Heat Transfer on a Non-Orthogonal Mesh Using the Finite-Volume Method. Numerical Heat Transfer, Part B, 23: 269-288, 1993.

14. Raithby, G.D., Chui, E.H. A Finite-Volume Method for Predicting a Radiant Heat Transfer in Enclosures with Participating Media. J. Heat Transfer 112: 415-423, 1990.
15. Hottel, H.C., Sarofim, A.F. Radiative Transfer. McGraw-Hill, New York, 1967.
16. Siegel, R., Howell, J.R. Thermal radiation heat transfer, $4^{th}$ edition. Taylor & Francis, 29 West $35^{th}$ street, New York, 2002.
17. Cumber, P.S., Fairweather, M. Evaluation of flame emission models combined with the discrete transfer method for combustion system simulation. International Journal of Heat and Mass Transfer 48: 5221-5239, 2005.
18. Soufiani, A., Taine, J. High temperature gas radiative property parameters of statistical narrow-band model for $H_2O$, $CO_2$ and CO, and correlated-K model for $H_2O$ and $CO_2$. International Journal of Heat and Mass Transfer 40: 987-991, 1997.
19. Modest, M.F. Radiative heat transfer. Academic Press, San Diego, 2003.
20. Andersson, K., Johansson, R., Johnsson, F., Leckner, B. Radiation intensity of propane-fired oxy-fuel flames: Implications for soot formation. Energy & Fuels 22: 1535-1541, 2008.
21. Shih, T.-H., Liou, W.W., Shabbir, A., Yang, Z., Zhu, J. A new k-ε eddy viscosity model for high Reynolds number turbulent flows. Computers and Fluids 24 (3): 227-238, 1995.
22. Bäckström, D., Jilvero, H. CFD-modelling of Chalmers 100 kW oxy-fuel unit – Investigation of swirling air and oxy-fuel flames. Report no. T2009-324. Chalmers University of Technology, Gothenburg, Sweden, 2009.
23. Jones, W.P., Lindstedt, R.P. Global reaction schemes for hydrocarbon combustion. Combustion and Flame 73: 233-249, 1988.
24. Magnussen, B.F. On the structure of turbulence and a generalized eddy dissipation concept for chemical reaction in turbulent flow. $19^{th}$ AIAA Aerospace meeting, St. Louis, 1981.
25. Pope, S.B. Computationally efficient implementation of combustion chemistry using in-situ adaptive tabulation. Combustion Theory and Modeling 1: 41-63, 1997.
26. Andersson, K. Characterization of oxy-fuel flames – their composition, temperature and radiation. Thesis for the degree of doctor of philosophy, Chalmers University of Technology, Gothenburg, Sweden, 2007.
27. Chui, E.H., Douglas, M.A., Tan, Y. Modeling of oxy-fuel combustion for a western Canadian sub-bituminous coal. Fuel 82: 1201-10, 2003.
28. Brink, A., Kilpinen, P., Hupa, M., Kjäldman, L. Study of alternative descriptions of methane oxidation for CFD modeling of turbulent combustors. Combustion Science and Technology 141: 59-81, 1999.
29. Peng, L., Zhang, J. Simulation of turbulent combustion and NO formation in a swirl combustor. Chemical Engineering Science 64(12): 2903-2914, 2009.

# Particle Image Velocimetry of Pulverized Oxy-Coal Flames

Husam el Gendy, Ignacio Preciado, Terry Ring and Eric Eddings

Department of Chemical Engineering, University of Utah, Salt Lake City, Utah 84112

## Introduction

Oxy-fuel combustion of pulverized coal is a promising technology for cost-effective power production with carbon capture and sequestration that has impacts on emission reductions. To fully understand the behavior of turbulent oxy-coal flames, and to validate oxy-coal simulation models, accurate experimental data are needed on the details of the flow field and dynamics of pulverized coal flames at multiple scales. A non-intrusive laser diagnostic technique, Particle image velocimetry (PIV), has been used in the past to analyze the velocity field of flames under various combustion conditions. In this technique a laser sheet, formed using a combination of spherical and cylindrical lenses, illuminates seeding particles in the flow at two different times and images are recorded on a camera that is synchronized with the laser. Using PIV software, the images are divided into interrogation areas and cross-correlated to determine the displacement between the two images. Knowing the time between the two laser pulses, velocity components and all velocity-derived quantities are calculated.

Although PIV has seen considerable application in laboratory flames, there has been limited application of this technique in pulverized-coal flames. In this work, PIV measurements were made of pulverized oxy-coal flames in a laboratory-scale laminar diffusion flame burner.

## Experimental set-up

Figure 1 shows the main components of the experimental set-up, which include a coal burner, a coal feeder and the PIV experimental setup.

The lab-scale burner utilizes four concentric streams to control the ignition and flame-shape. The coal is transported from the coal feeder by a carrier gas (air) through the central tube of the burner. Immediately circling the central coal jet there is a natural gas pilot whose purpose is to ignite the coal. Outside of the pilot ring there is an air annulus that will provide air to both the pilot and coal flames and will allow for control of the overall stoichiometric ratio during coal combustion. The final annular ring (outside of the air stream) is a pre-mixed ethylene/air stream.

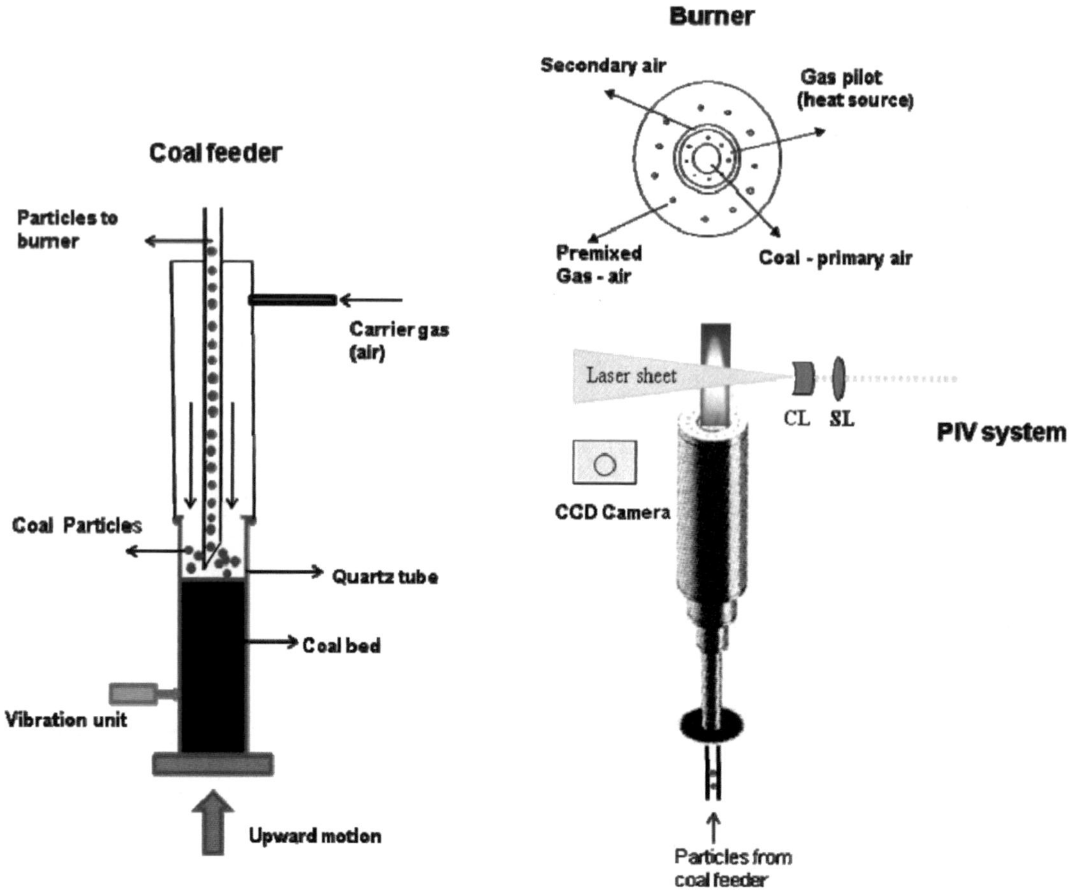


**Fig. 1.** Lab-scale burner, coal feeder and PIV experimental setup.

The primary purposes of this stream are to heat the air stream, facilitating ignition of the coal jet, and to reduce heat loss from the flame.

The design concept for the coal feeder is a motor-driven unit that constantly advances a powdered coal surface toward a jet of air that fluidizes the coal and carries it out of the feeder to the central tube of the burner.

A quartz tube (0.95 cm ID, 1.27 cm OD and 19 cm long) is used to store the coal and is supported by a metal plate, which is driven upwards with the motor. A stainless steel tube (2.2 mm ID and 3.2 mm OD) is placed at a fixed position and put inside the quartz tube, and air is used as carrier gas to transport the coal particles. The coal is fluidized and transported upwards by the carrier gas through the stainless steel tube, which is connected to the burner. A vibration unit is in contact with the quartz tube in order to prevent agglomeration of the coal and facilitate particle fluidization and transport. The motor moves the quartz tube upwards at the same speed as the coal level decreases, keeping constant the distance between the coal-bed surface and the inlet of the stainless steel tube. In this way, the mass flow rate of coal fed to the burner is uniform.

Variables such as motor speed (it determines the vertical speed at which the coal feeding system moves upwards), volumetric flow rate of carrier gas and the intensity of the vibration unit determine the coal flow rate given by the feeder.

The PIV system consists of a double frame CCD camera (Imager Pro X, 12 bits, 2048 x 2048 format, Lavision), a double headed laser (Nd:YAG, 532 nm, 50 mJ/pulse, 7.26 Hz in double mode, Lavision). The camera and the lasers are synchronized through a programmable timing unit (PTU 7, Lavision). The laser beams are transformed into a laser sheet, of thickness ~ 1mm, using spherical and cylindrical lenses of focal lengths of 500 and -200 mm, respectively. To minimize the effect of flame luminosity an interference filter centered at the 532 ± 5 nm (Edmund) was used and mounted in the camera lens (Tamron). Also a camera shutter (Lavision) was mounted in the camera lens to shorten the exposure time of the $2^{nd}$ frame taken by the camera from 67 ms to 3 ms. Without this shutter it is impossible to get PIV information as the $2^{nd}$ frame is always over exposed with the luminosity of the flame due to the long time of the exposure time of the $2^{nd}$ frame. The images are cross-correlated using the PIV software (Davis 7, Lavision) with an initial interrogation size of 128 px x 128 px for initial pass and 64 px x 64 px for final pass. Each frame is 3.8 cm x 3.8 cm with a resolution 18 µm/px. The time between laser pulses, 50 µs, and the size of the interrogation window, 64 px x 64 px (1.2 mm x 1.2 mm), were chosen such that the maximum displacement from frame a to frame b is less than ¼ of the interrogation area size, 16 Px.

The flame height is about 15 cm and a vertical stage was used to move the burner assembly up and down to capture the pulverized coal flame at three different heights. Each PIV experiment is repeated three times; one for the part of the flame starting from the burner's head to 3.5 cm vertically, then from 3.5 cm to 7 cm and finally from 7.5 cm to 10.5 cm. There was no seeding particles used in these experiments. The pulverized coal particles served as the seeding particles themselves. Each experimental data set consists of 200 consecutive double frames taken within ~1 minute.

The results presented in this paper were obtained for a selected condition of mass flow rate of coal (100 mg/min), and for a volumetric flow rate of carrier gas of 1.8 L/min. The variation of the coal flow rate was less than 13%. Coal particle velocities were estimated under different global equivalence ratios (ϕ) of the system (ranging from 0.25 to 0.92) and for different coal particle sizes.

**Results and discussion:**

**PIV of pulverized coal flames:**

Fig. 2 shows a sample of an instantaneous Mie scattered images of pulverized coal flames and the corresponding velocity fields at 3 different heights of the flame with $\phi = 0.92$ and using coal particles of sizes < 38 μm. It is to be noted that each image is one of twin frames taken at two different times, t and t + dt (dt = 50 μs). From these twin frames one velocity map is created. The inner cone in the flame is formed due to the premixed coal-air stream fed through the central tube of the burner. In the images the burner head is seen at the bottom where the outer diameter of the central part is 7.2 mm. As can be seen in Fig.2, the velocity fields of pulverized coal flames show that the lower parts of the flame are more stable than the upper parts and as such have lower axial velocities than upper parts. To show the dynamic nature of the coal flame as we move vertically up in the flame, the average instantaneous velocity inside a small area in the velocity field on the flame axis are compared at three different heights (Fig. 3 left ). It is apparent that time-average mean velocity and the fluctuation in the instantaneous velocity during the 1 minute run increase vertically. Another important consequence of this increased fluctuation in the upper parts of the flame is that it happens that the flame may move perpendicular to the laser sheet during the run and therefore no coal particles are present to scatter the laser and the image becomes blank. This results in some zero velocities in the middle and upper flame parts- heights 5 and 8.5 cm. This increase in velocity is a result of the acceleration of the flow vertically due to air entrainment. This fluctuation looks random as no specific periodicity was observed in the power spectrum as shown in the right plot of Fig.3. Also the standard deviation was found to increase slightly with height above the burner head.

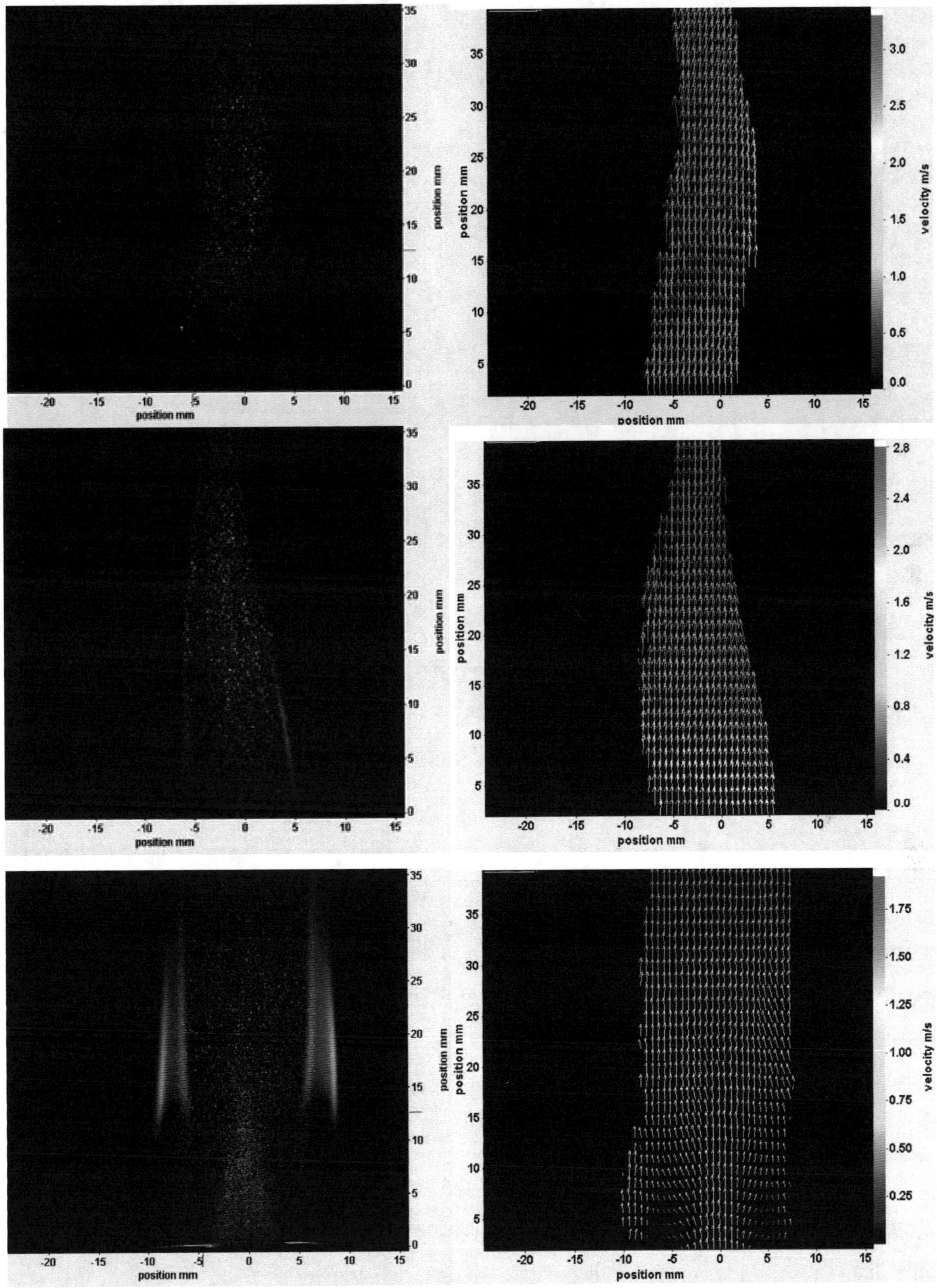


**Fig. 2.** Image maps (left), and the corresponding velocity fields (right) of pulverized coal flame. Viewing the flame from burner head to a height 3.5 cm (bottom,), 3.5-7 cm (middle), and 7-10.5 cm (top). $\phi = 0.92$ and coal particle sizes < 38 $\mu$m.

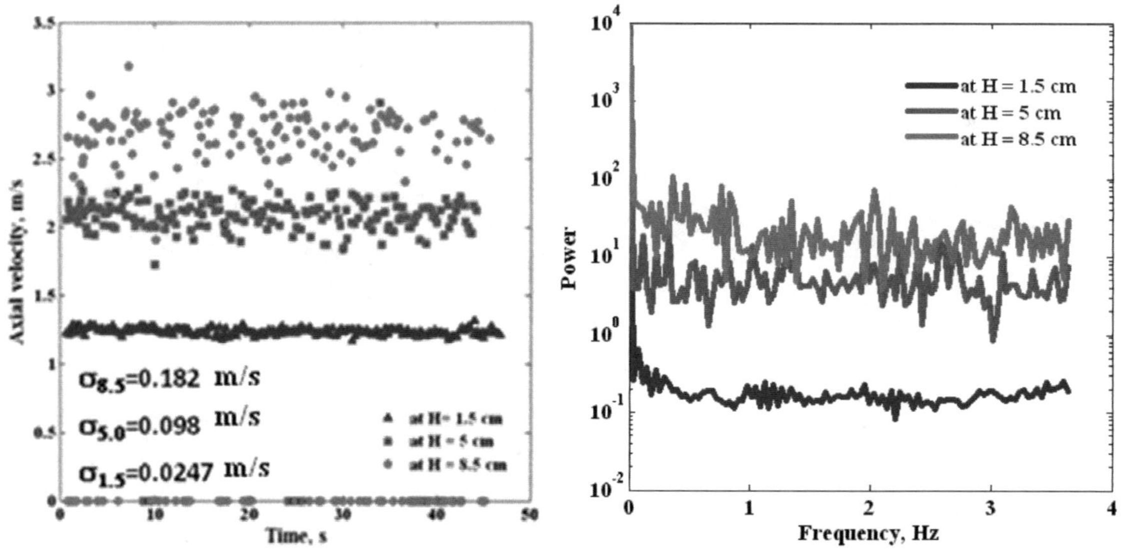

**Fig. 3.** Fluctuation in axial velocity at 3 different heights (left plot) and corresponding power spectrum (right plot). Coal particle size < 38 μm and $\phi = 0.25$. $\sigma$ is the standard deviation.

**Effect of equivalence ratio and coal particle size:**

Pulverized coal particles of different size bins were prepared in order to investigate the effect of coal particle size on the velocity field and the dynamics of the combustion of pulverized coal. Additionally, experiments were carried out under three different equivalence ratios ($\phi$ = 0.25, 0.44 and 0.92). Fig.4 shows that as the equivalence ratio of the pulverized coal flame increases, the height of the inner cone in the pulverized coal flames decreases.[1] In these experiments increasing the equivalence ratio is made by decreasing the mass flow rate of the secondary air (air fed through the annular ring). This reduction in the air flow rates results in less air entrainment in the inner cone of the flame and hence a reduction in the cone height. Furthermore, inspecting the instantaneous images in Fig. 4 it is clear that with high excess air (very lean fuel-air mixture, $\phi$= 0.25) burning coal particles can be identified more easily than with higher $\phi$. For all coal particle sizes it was found that increasing $\phi$ results in an increase in the axial velocity. This trend holds true inside and outside the inner cone (see Fig. 5).

It is evident from Fig. 5 that flames with coarser coal particles have, in general, less axial velocity than those with smaller coal particle sizes. Also the rate of the reduction in the axial velocity increases as the equivalence ratio decreases as mentioned before, although the trend varies somewhat as stoichiometric conditions are approached. Although there is some scatter observed in the data, there may be trade-offs between hydrodynamic effects at low vs. higher

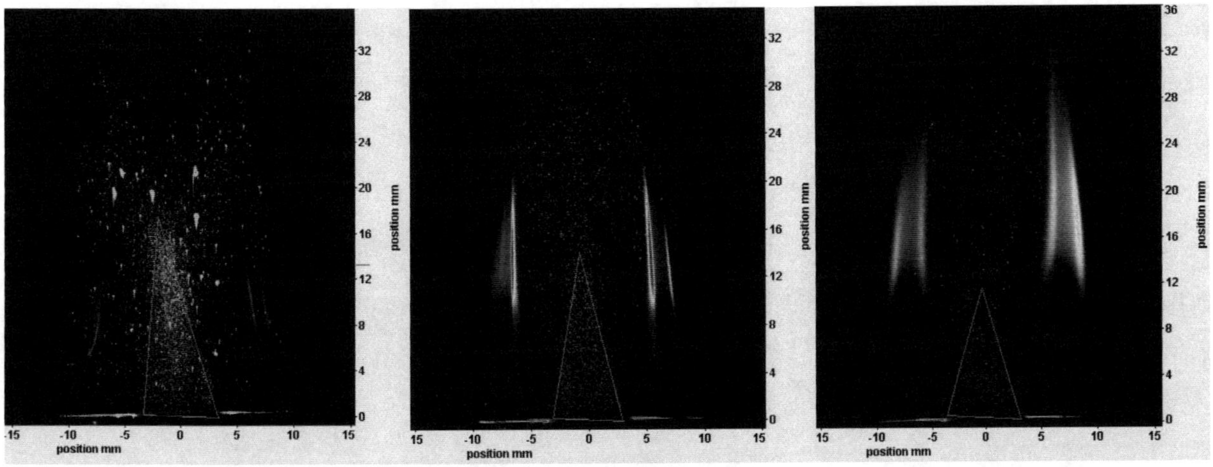


Φ= 0.25 Φ= 0.44 Φ=0.92

**Fig. 4.** PIV Images of pulverized coal flames at different ϕ. Coal particle size is < 38 μm.

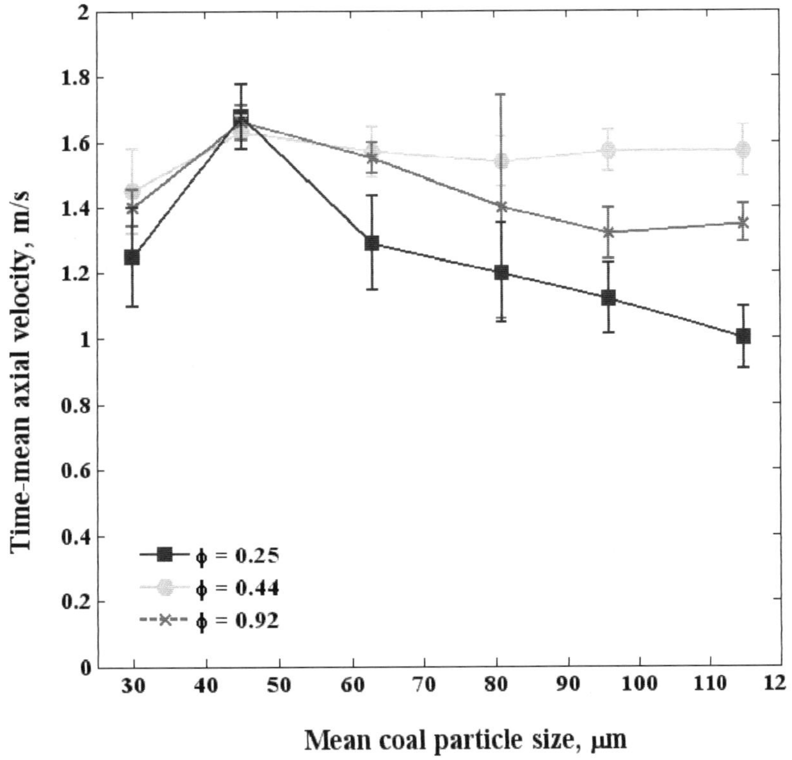


**Fig. 5.** Mean axial velocity at height 3 cm above burner head using coal particles of different mean particle sizes and at three different equivalence ratios, ϕ.

equivalence ratios, and temperature effects due to particle ignition, which may be impacted by both local oxygen concentration and particle size. Larger coal particles will have increased drag and greater terminal velocities in this upwardly-fired flame, and thus would tend to have greater slippage with the surrounding gas velocity. Smaller particles will more likely track the gas velocity and exhibit much less difference between gas and particle velocities. Additional experiments are planned to separately measure the gas velocity to quantify the gas-solid velocity differences. Since larger particles will also take a longer time to devolatilize[2], and given the significant differences in local oxygen concentration between the equivalence ratios utilized, the ignition behavior and thus temperature could also be quite different for different particle sizes and different equivalence ratios. These trade-offs between hydrodynamic effects and ignition may affect an anticipated trend with equivalence ratio.

**Conclusions**

- PIV was successful in analyzing laminar coal flames using the coal particles themselves as seeding particles.
- It was found that the mean axial velocity increases vertically in the flame, and lower portions of the flame are more stable than upper portions of the flame.
- Regardless of the fluctuation of the flame, especially in upper portions of the flame, there was no specific periodicity over the PIV run time, as detected by the power spectrum of velocity.
- The mean axial velocity is increasing as the equivalence ratio increases.
- Increasing the coal particle size in the flame results in a decrease in the mean axial velocity and the rate of decrease in the velocity decreases with increasing equivalence ratio.

**References**


[1] D. F. Kurtuluş, et al., "Characterisation of Lean Premixed Laminar Flames in High Pressure using PIV", Proceedings of the European Combustion Meeting 2009.

[2] Y. Yan, et al., "Monitoring and characteristics of pulverized coal flames using digital imaging techniques", Fuel 81 (2002) 647-656.

# (552d) The Relative Contribution of Particle and Gaseous Radiation in Oxy-Fuel Combustion

**Robert Johansson**, Klas Andersson and Filip Johnsson, Energy and Environment, Chalmers University of Technology, Goeteborg, Sweden

Radiation is the main mechanism for heat transfer in a combustion chamber and it is therefore essential to have a good understanding of it for the design of heat transfer surfaces. The changed conditions of oxy-fired, compared to air-fired boilers, have important effects on the radiative characteristics of the combustion gas. The increased concentrations of carbon dioxide and/or water vapor, depending on whether dry or wet flue gases are recycled, results in an enhanced contribution to emission and absorption of thermal radiation. The impact of these changes depends not only on increased molar fractions of the gaseous species, but also on the relative importance of gaseous radiation compared to particle radiation and how these two interact. In a coal-fired boiler, particles in the form of fuel, soot and ash contribute to the emitted radiation and can in the flame zone dominate the radiative heat transfer. The radiative properties of particles are characterized by relatively small spectral variations and can to a large extent be treated as gray, while the properties of the combustion gases show a complex spectral behavior, which has significant effects on the radiative properties. Due to these differences there is a need to investigate both gas and particle radiation in oxy-fired atmospheres, to be able to quantify changes compared to air-fired units. Such knowledge is of large value in the scaling of the oxy-fuel process, in which boilers of industrial scale will be designed, based on studies in lab and pilot scales and previous experience of air-fired facilities.

This work focuses on radiation in oxy-fired conditions. Both gas and particle radiation is modeled in an axi-symmetric cross section of a cylinder and differences between air- and oxy-firing are investigated. The intensity field is calculated according to the discrete transfer method by solving the intensity along a number of rays that crosses a radial profile. Scattering along the rays is related to the distance to the centre of the cylinder. For the gas radiation, a Statistical-Narrow-Band (SNB) model is applied as reference while the particle radiation is modeled by empirical correlations accounting for spectral properties of coal particles. Scattering of the particles is assumed to be isotropic. Included in the analysis is also a non-gray or banded formulation of the Weighted-Sum-of-Gray-Gases (WSGG) model and grey approximations. These two approaches are computationally efficient and reasonable to use in CFD-simulations. A grey approximation is the most simplified approach and probably the most common way to handle gas radiation in CFD-modeling of combustion. The banded formulation of the WSGG model is an effective method to increase the accuracy of the radiative source term at a relatively low computational cost.

The investigated cases cover both air-and oxy-fired conditions and the properties of the combustion gas are based on measured data from a lignite flame in Chalmers 100 kW rig. Temperatures spans from 1000 K to 1800 K and gas concentrations corresponding to both dry and wet flue gas recycling are examined. The particle load is obtained by fitting modeled radiation intensity profiles against intensity profiles measured by a narrow angle radiometer. Wall fluxes and the radiative source term along the cylinder diameter are compared to evaluate differences in the radiative heat transfer between air- and oxy-fuel combustion. Included in the study is also a sensitivity analysis of the influence of particle load and scattering. The different approximations for gaseous radiation are evaluated to quantify the errors they result in when they are applied in conditions with a high particle load. Scale effect are also addressed by increasing the diameter of the cylinder while maintaining the same profiles of temperature, gas species and particle concentration.

See more of this Session: Oxycombustion of Coal - Need, Opportunities, and Challenges III
See more of this Group/Topical: Topical E: High Temperature Environmentally Sustainable Energy Processes (sessions joint with the Environmental Division)

# (552e) Experimental Studies of Ignition in 40kW Co-Axial Turbulent Diffusion Oxy-Coal Jet Flames

**Jingwei Zhang**[1], Kerry Kelly[2], Eric G. Eddings[1] and Jost O. L. Wendt[1], (1)Chemical Engineering, University of Utah, Salt Lake City, UT, (2)Institute for Clean and Secured Energy, University of Utah, Salt Lake City, UT

This paper is directed towards understanding how composition of the coal transport medium (primary CO2 and O2) affects the ignition stand-off distance in 40kW co-axial turbulent diffusion, oxy-coal flames, supported in a specially designed combustion test rig, rated at 100kW. First, a methodology to define and quantify the ignition behavior of laboratory-combustor-scale turbulent pulverized coal jets was developed and is described in detail. Results are presented in the form of probability density profiles of the measured stand-off distance, as obtained from replicate runs each consisting of six thousand photo images. Multimodal behavior of the PDFs describing flame standoff distances was observed. Measured mean standoff distance and predicted ignition time were correlated. Results show that under turbulent mixing, the controlling mechanism of affecting coal jet ignition and flame stability is the molecular diffusion of O2 to the coal particle surface. Not only do these results provide qualitative insight into ignition of coal in turbulent oxy-coal flames of practical relevance, but they can also provide a basis for validation of future detailed simulations of this process.

See more of this Session: Oxycombustion of Coal - Need, Opportunities, and Challenges III
See more of this Group/Topical: Topical E: High Temperature Environmentally Sustainable Energy Processes (sessions joint with the Environmental Division)

# (552f) The EFFECT of COAL Composition ON Ignition and Flame Stability IN CO-Axial Oxy-FUEL Turbulent Diffusion Flames

**Dadmehr M. Rezaei**[1], Eric G. Eddings[1], Kerry E. Kelly[1], Jingwei Zhang[1], Yuegui Zhou[2] and Jost O.L. Wendt[1], (1)Chemical Engineering, University of Utah, Salt Lake City, UT, (2)Shanghai Jiao Tong University, University of Utah, Salt Lake City, UT

Abstract:

Past research on flame stability and stand-off distance under oxy-coal combustion conditions has used a 100kW pulverized coal test rig with a co-axial turbulent diffusion burner, and has been described at previous Pittsburgh Coal Conferences. These studies were for one specific coal, namely a Utah Bituminous Coal. The purpose of the research described in this paper is to extend the previous work and to explore how coal composition changes affect the following dependencies that control flame stand-off distance and flame ignition, namely: 1) the effect of partial pressure of oxygen (PO2) in the primary stream with differing preheat temperatures in the secondary stream; and 2) the effect of PO2 in the secondary stream with zero O2 in the primary stream. The results of this new study were designed to extend previously obtained knowledge on effects of secondary preheat temperature, turbulent mixing, PO2 in various streams, from one single coal to other coals of differing compositions.

This paper, therefore, explores the effects of coal composition on ignition in oxy-coal, coaxial, turbulent diffusion flames. In this research, the stability and stand-off distance of the flame were studied for the following three types of coal: Utah Skyline Bituminous, Illinois #6 Bituminous, and a Powder River Basin (Black Thunder) coal. To this end we investigated: 1) the effect of PO2 in the primary stream, 2) the effect of PO2 in the secondary stream, and 3) the effect of preheat temperature in the secondary stream, on flame stand-off distance, using the same photo-imaging methodology described elsewhere. The results of the ignition and flame stability analysis for these three coals under oxy-firing conditions are compared, and the effects of coal composition are elucidated.

See more of this Session: Oxycombustion of Coal - Need, Opportunities, and Challenges III
See more of this Group/Topical: Topical E: High Temperature Environmentally Sustainable Energy Processes (sessions joint with the Environmental Division)

# (552g) Oxy-Gas Combustion for Process Heater Retrofitting: Effect of Scenario Parameters On Velocity and Composition Fields in the near Burner Region

Jennifer Spinti[1], Jeremy N. Thornock[2] and Philip J. Smith[1], (1)Chemical Engineering, University of Utah, Salt Lake City, UT, (2)University of Utah, Salt Lake City, UT

Oxy-fuel combustion has been proposed as a retrofit technology for the power generation, petrochemical, and refining industries. Despite its advantage of providing an effluent stream of concentrated CO2, questions remain about heat transfer, pollutant formation, and flame stability in boilers, process heaters, and other combustion equipment retrofitted with oxy-fuel burners.

The objective of our oxy-gas research is to produce the simulation tools needed for simulating efficient CO2 capture from process heaters. We have specifically focused on (1) creating a Large Eddy Simulation (LES) tool for demonstrating practical oxy-gas combustion in process heaters and (2) using the LES technology to produce predictive capability with quantified uncertainty bounds for a pilot-scale oxy-gas process heater.

We use a rigorous uncertainty quantification (V/UQ) analysis approach to achieve predictive capability, employing the Data Collaboration methods of Michael Frenklach and coworkers at the University of California-Berkeley [1,2]. Data Collaboration requires consistency between simulation results and experimental data. For our Data Collaboration analysis, we employ the ARCHES simulation tool, a three-dimensional, LES code developed by Professor Smith and his research group at the University of Utah. ARCHES uses a low-Mach number ($M < 0.3$), variable density formulation to simulate heat and mass transfer in reacting flow [3]. The selected experimental data are from the International Flame Research Foundation's (IFRF) oxy-gas experiments, also known as the OXYFLAM experiments [4].

To perform the V/UQ analysis, there are requirements of both the experimental data and the simulations. For the experimental data, the uncertainty range of all measured variables (response quantities of interest) must be determined as precisely as possible. For the simulations, the parameter space that has the greatest effect on the experimentally measured variables must be probed via an experimental design. In the case of the IFRF OXYFLAM experiments,the parameter space that we have probed includes scenario parameters such as O2 and natural gas flow rates, model parameters such as the model for the boundaries of the near burner region, and numerical parameters such as mesh resolution. We examine the effect of the ranges of these parameters on the simulation results and determine the region of consistency between experiment and simulation. This region of consistency defines the uncertainty range of the response quantities of interest.

References [1] Frenklach, M., Packard, A., Seiler, P., and Feeley, R., "Collaborative Date Processing in Developing Predictive Models of Complex Reaction Systems," International Journal of Chemical Kinetics, 36, 57–66 (2004). [2] Feeley, R., Seiler, P., Packard, A., and Frenklach, M., "Consistency of a Reaction Dataset," Journal of Physical Chemistry A, 108, 9573–9583 (2004). [3] Spinti, J., Thornock, J., Eddings, E., Smith, P., and Sarofim, A., "Heat Transfer to Objects in Pool Fires," in Transport Phenomena in Fires, WIT Press, Southampton, U.K., 2008. [4] Lallemant, N., Dugue, J., and Weber, R., "Analysis of the Experimental Data Collected During the OXYFLAM-1 and OXYFLAM-2 Experiment, Phase 1: 1995-1996. IFRF Doc No F 85/y/4, International Flame Research Foundation, IJmuiden, The Netherlands (1997).

See more of this Session: Oxycombustion of Coal - Need, Opportunities, and Challenges III
See more of this Group/Topical: Topical E: High Temperature Environmentally Sustainable Energy Processes (sessions joint with the Environmental Division)

# (574a) Pressurized Coal Pyrolysis and CO2 Gasification at High Initial Heating Rates

Randy Shurtz, Sam Goodrich, Greg Sorensen and Thomas H. Fletcher, Chemical Engineering, Brigham Young University, Provo, UT

An understanding of the effects of pressure and initial heating rates on coal pyrolysis is critical to modeling gasification because the char structure can have a significant effect on apparent reaction rates. Previous studies at atmospheric pressure have shown that coal swelling reaches a maximum at heating rates of ~$10^4$ K/s (the operating regime of conventional drop-tube reactors) and then decreases to an asymptotic minimum at ~$10^5$ K/s (the operating regime of flat-flame burners). It is believed that the swelling behavior observed at $10^5$ K/s is more representative of industrial gasifiers, which operate at ~$10^6$ K/s. The BYU High Pressure Flat-Flame Burner (HPFFB) has been rebuilt to be more suitable for gasification studies at heating rates of ~$10^5$ K/s and pressures of 2.5 atm - 15 atm. The HPFFB has been used to conduct pyrolysis and gasification experiments on three bituminous coals and a subbituminous coal. Swelling of bituminous coals at high initial heating rates ($10^5$ K/s) was found to exceed swelling from similar experiments at atmospheric pressure, but the measured swelling ratios were low compared to values reported from pressurized experiments conducted at lower heating rates ($10^4$ K/s). Subbituminous coals produced highly porous chars with high reactivities. Swelling information from these tests has been combined with published data to develop a swelling correlation with heating rate dependence for CFD applications. Mass release information gained from these tests has been used to fit kinetic coefficients in a gasification model.

See more of this Session: Advances in Gasification Research I
See more of this Group/Topical: Topical E: High Temperature Environmentally Sustainable Energy Processes (sessions joint with the Environmental Division)

# (574b) Pyrolysis and Gasification Reactivity of Coal and Petcoke Under High Pressure Conditions

**David Ray Wagner**[1], Kevin J. Whitty[2], Paul Fanning[3] and Glenn Shoaf[3], (1)Institute for Clean & Secure Energy, The University of Utah, Salt Lake City, UT, (2)Institute for Clean & Secure Energy, Dept. of Chemical Engineering, The University of Utah, Salt Lake City, UT, (3)Gasification Process Development, Eastman Chemical Company, Kingsport, TN

Nearly all gasification systems targeted for use in IGCC or chemical/fuel production systems operate at high pressures, sometimes exceeding 1000 psi. In order to better understand the processes by which gasification feedstocks are converted in high pressure systems, three fuels--a lignite coal, an eastern bituminous coal and petroleum coke--were subjected to a series of laboratory experiments. The objective of the study was to evaluate pyrolysis behavior and reactivity under high pressure conditions relevant for entrained-flow gasification applications. Pyrolysis tests were conducted in nitrogen using a high pressure wire mesh heater, which allowed fast heating rates (in excess of 1000 K/s) and pressures to 900 psi. Reactivity of the fuels was studied by thermogravimetric analysis under gasification conditions. This paper compares devolatilization behavior (e.g., volatiles and char yields) of the fuels under a variety of pressures and final temperatures. Gasification rates of chars formed from the fuels are also presented.

See more of this Session: Advances in Gasification Research I
See more of this Group/Topical: Topical E: High Temperature Environmentally Sustainable Energy Processes (sessions joint with the Environmental Division)

# The Effects of Temperature Upon the Supercritical Water Reformation of Alcohols for Hydrogen Production

*Jared Bouquet[1], Jason Picou[1], Michael Stever[1], Jonathan Wenzel[2], Ryan Tschannen[3], and Sunggyu Lee[3]*

*1. Missouri University of Science and Technology, Rolla, MO*
*2. Kettering University, Flint, MI*
*3. Ohio University, Athens, OH*

## *Abstract*

Methanol is a very common industrial chemical used in a variety of processes, one of which is the transesterification of triglycerides to make biodiesel. The crude glycerol byproduct of the transesterification reaction contains a substantial amount of methanol. Supercritical water reformation of crude glycerol presents a unique non-catalytic means of converting unwanted crude glycerol into hydrogen for use in energy applications, hence increasing the amount of usable transportation fuel that may be produced from triglycerides in biodiesel production. The effect of reformation on each of the individual components of crude glycerol must be studied individually so that a detailed analysis on the effects of reformation of crude glycerol as a whole can be performed. The effect of temperature was evaluated on the non-catalytic reformation of methanol in supercritical water. An experimental study was conducted using a 0.1-L Inconel 625 reactor at a constant pressure of 24.2 MPa and constant flow rate of 8 grams per minute. The temperature was varied between 500 and 650°C with water to methanol molar ratios ranging from 1:1 to 13:3.

## *Introduction*

Methanol is a very common industrial chemical used in a variety of applications. One of those applications is as a reactant for the production of biodiesel. In the biodiesel production process, three molecules of methanol react with a triglyceride to make three molecules of biodiesel and a single molecule of glycerin. Once the biodiesel is removed from the resulting mixture, the remaining solution is known as crude glycerol and is comprised of mostly glycerin and excess methanol reactant.

Biodiesel is an alternative liquid transportation fuel that has the potential to readily displace its petroleum counterpart, since it can be used in conventional diesel engines with little to no modifications.[1] With the emergence of biodiesel as an alternative fuel, finding means to utilize the crude glycerol byproduct has become an area of interest. The reason for this is that more glycerol is being created than can be used commercially in conventional enduses.[2]

Previous work by Picou et al. has explored the supercritical water reformation of pure glycerin for hydrogen production.[3] Before exploring the effect of supercritical water reformation on crude glycerol for hydrogen production, the effect of supercritical reformation on pure methanol should be explored. The results of the pure methanol study will help give insight into the reformation of a crude glycerin solution. For this study hydrogen was produced from methanol by a novel process, non-catalytic supercritical water reformation. In order to determine the effect of temperature and water-to-methanol ratio on the supercritical water reformation of methanol, a series of experiments were conducted at 24.2 MPa and a flow rate of 8 grams per minute at temperatures between 500 and 650°C and water-to-methanol molar ratios ranging from 1:1 to 13:3.

### *Chemical Reactions*

There are many chemical reactions that occur during when methanol is exposed to supercritical water. The principal reactions discussed in this paper are methanol decomposition, water-gas shift reaction, supercritical water reformation, and methanation reactions. Equation 1 shows the chemical reaction for methanol decomposition.

**Equation 1 Methanol Decomposition:** $CH_3OH \leftrightarrow 2H_2 + CO$

This reaction is endothermic according to N. Boukis et al. with a heat of reaction of +91.7 kJ per mol.[4] This reaction produces the bulk of hydrogen for these experiments, making two moles of hydrogen for every mole of methanol. The other reaction that makes hydrogen is the water-gas shift reaction, which is shown in Equation 2.

**Equation 2 Water-Gas Shift:** $CO + H_2O \leftrightarrow CO_2 + H_2$

The water-gas shift (WGS) reaction is exothermic with a heat of reaction of -41 kJ per mol.[4] The reaction is reversible in nature, and involves the combination of water and carbon monoxide to make hydrogen and carbon dioxide. The forward reaction is thermodynamically favored at temperatures of 815°C or below, which is true for all of the experiments discussed in this paper.[5] Assuming that all carbon monoxide produced in the decomposition reaction subsequently undergoes the WGS reaction, the overall reaction would be that of supercritical water reforming of methanol, which is shown in Equation 3.

**Equation 3 Supercritical Water Reformation of Methanol:** $CH_3OH + H_2O \leftrightarrow CO_2 + 3H_2$

The supercritical water reformation equation is basically Equations 1 and 2 combined, and as such it has a heat of reaction is equal to the sum of the two heat of reactions, which results in +50.7 kJ per mol. This Equation 3 also shows that the theoretical maximum of hydrogen produced is three moles of hydrogen for every mole of methanol fed. Optimally all methanol fed would undergo reformation, resulting in only hydrogen and carbon dioxide in the effluent. Since carbon monoxide is present in the experimental runs, it can be assumed that the water-gas shift reaction is incomplete. Also the presence of methane indicates that methanation is occurring, which is show in Equations 4 and 5

**Equation 4 Methanation of Carbon Monoxide:** $CO + 3H_2 \leftrightarrow CH_4 + H_2O$

**Equation 5 Methanation of Carbon Dioxide:** $CO_2 + 4H_2 \leftrightarrow CH_4 + 2H_2O$

The methanation reaction is a reversible reaction that involves carbon oxides and hydrogen reacting to become methane and water. The reactions are both exothermic; with heats of reaction of -211 kJ per mol for carbon monoxide methanation and -223 kJ per mol for carbon dioxide reformation.[4] The forward methanation reaction consumes hydrogen, and therefore is detrimental to the reformation process.

### *Apparatus and Chemicals*

The chemicals used for the experiments were deionized water and methanol. The water and methanol for the experiments were then mixed into solutions with concentrations between 29.1 wt-% and 64.0 wt-% methanol. These solutions correspond to a water-to-methanol molar ratio of 13:3 and 1:1 respectively. The physical properties of methanol and water may be found in Table 1.

**Table 1.** Physical Properties of Methanol and Water[6,7]

| Property | Methanol | Water |
|---|---|---|
| Molecular Weight (g/mol) | 32.04 | 18.02 |
| Density at 25 ˚C and 1 bar (g/cm$^3$) | 0.79 | 0.99 |
| Melting Point (˚C) | -98 | 0 |
| Boiling Point (˚C) | 64 | 100 |
| Critical Temperature (˚C) | 512 | 647 |
| Critical Pressure (atm) | 79.7 | 217.7 |

The supercritical water reformation system used for these experiments consists of a water and a fuel feed system, reactant preheater, Inconel® Alloy 625 reactor with a heater, liquid collection system, gas sampling system, and a data acquisition and control system. Figure 1 is a representative process flow diagram of the supercritical water reaction system. The Inconel® 625 reactor has an internal volume of 101 mL, and is an alloy of mostly nickel, chromium, iron, and molybdenum, along with other elements.[8] The samples collected of the gaseous effluent at ambient conditions were analyzed using a HP 5890 Series A gas chromatograph with a thermal conductivity detector. The gas chromatograph is calibrated to detect hydrogen, nitrogen, carbon monoxide, methane, carbon dioxide, acetylene, ethylene and ethane.

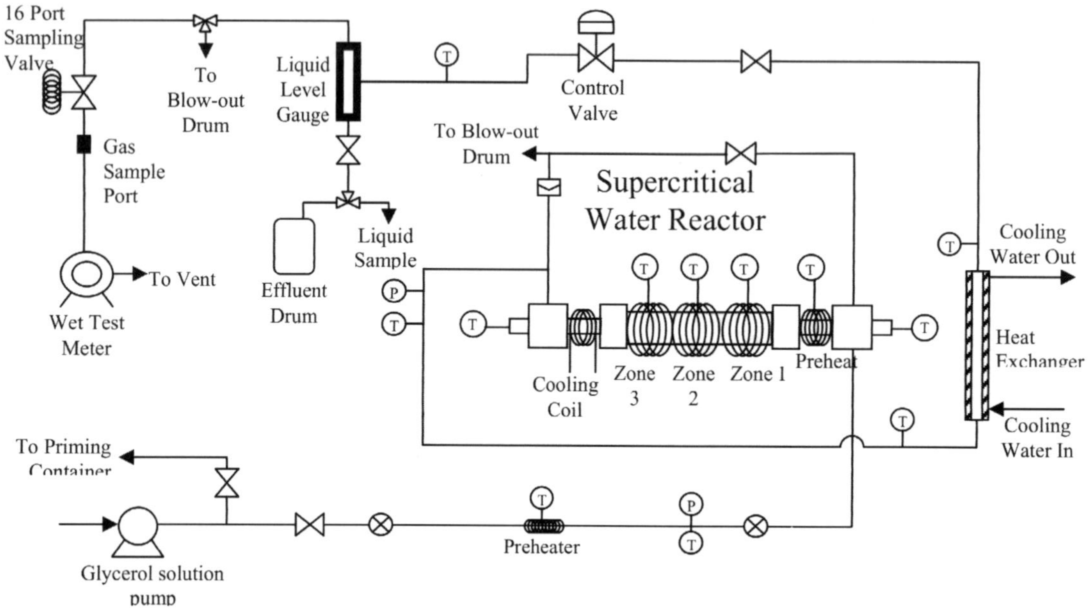


**Figure 1 The multi-fuel reformer system at the Missouri University of Science and Technology.**

## *Experimental*

The temperature and molar ratio of the water-methanol solution were varied for these experiments to determine their effect upon gas composition as well as the extent of the gasification of methanol. The experiments were performed non-catalytically at a constant pressure of 24.13 ± 0.2 MPa and a constant flow rate of 8.2 ± 0.6 grams per minute. The temperatures used in these experiments were 500 ± 2°C, 550 ± 1°C, 600 ± 1°C and 650 ± 2°C and water to methanol molar ratios of 1:1, 8:3, and 13:3. These specific molar ratios were chosen to have the same water-to-carbon molar ratios used in previous glycerin experiments. The space time was determined using the Peng-Robinson equation of state with Van der Waals mixing rules, and was determined to be approximately 60 seconds for all of the experimental runs. Table 2 lists the temperature, flow rate, pressure, water/methanol (W/M) ratio, and methanol weight percent of each experimental run.

**Table 2 Experimental Run Summary**

| Run No. | Temp (oC) | Pressure (psig) | Flowrate (g/min) | W/M ratio | MeOH wt% |
|---|---|---|---|---|---|
| WM-73M | 502 | 3506 | 8.1 | 1.0 | 64.0 |
| WM-74M | 551 | 3503 | 8.1 | 1.0 | 64.0 |
| WM-72M | 601 | 3507 | 8.1 | 1.0 | 64.0 |
| WM-75M | 652 | 3492 | 8.0 | 1.0 | 64.0 |
| WM-79M | 502 | 3505 | 8.2 | 2.7 | 40.0 |
| WM-78M | 550 | 3504 | 8.3 | 2.7 | 40.0 |
| WM-77M | 601 | 3505 | 8.2 | 2.7 | 40.0 |
| WM-76M | 651 | 3496 | 8.1 | 2.7 | 40.0 |
| WM-83M | 499 | 3471 | 8.3 | 4.3 | 29.1 |
| WM-82M | 551 | 3508 | 8.3 | 4.3 | 29.1 |
| WM-81M | 601 | 3477 | 8.1 | 4.3 | 29.1 |
| WM-80M | 651 | 3508 | 8.8 | 4.3 | 29.1 |

## *Results and Discussion*

The results of the experiments listed in **Error! Reference source not found.Error! Reference source not found.** are shown below in Table 3 and include the yield of the different gaseous species leaving the reactor relative to the amount of methanol fed and the total carbon conversion into gaseous species. Carbon not present in the gaseous effluent is either present in the liquid effluent, or remains in the reactor as a solid. The reactor was cleaned of any solid carbon residue either manually or by supercritical water oxidation prior to an experiment. While two of the experimental runs showed trace amounts of ethane, it is believed that it was from residual fuel in the system from previous experimental runs with different fuels, and should be disregarded for the analysis of results.

**Table 3 Gas Effluent Flowrates and Carbon Conversion**

| Run No. | Mol Gas Out/ Mol Methanol Fed | | | | | | | Carbon conversion |
|---|---|---|---|---|---|---|---|---|
| | $H_2$ | $N_2/O_2$ | CO | $CH_4$ | $CO_2$ | $C_2H_4$ | $C_2H_6$ | |
| WM-73M | 0.27 | 0.00 | 0.07 | 0.00 | 0.03 | 0.00 | 0.01 | 12.7% |
| WM-74M | 0.50 | 0.00 | 0.13 | 0.01 | 0.07 | 0.00 | 0.00 | 21.8% |
| WM-72M | 1.22 | 0.00 | 0.31 | 0.06 | 0.18 | 0.00 | 0.01 | 58.6% |
| WM-75M | 1.72 | 0.00 | 0.40 | 0.22 | 0.34 | 0.00 | 0.00 | 98.6% |
| WM-79M | 0.54 | 0.00 | 0.12 | 0.00 | 0.08 | 0.00 | 0.00 | 21.6% |
| WM-78M | 0.87 | 0.00 | 0.19 | 0.01 | 0.14 | 0.00 | 0.00 | 35.2% |
| WM-77M | 1.34 | 0.00 | 0.25 | 0.03 | 0.26 | 0.00 | 0.00 | 55.4% |
| WM-76M | 1.97 | 0.00 | 0.30 | 0.13 | 0.43 | 0.00 | 0.00 | 87.8% |
| WM-83M | 0.66 | 0.00 | 0.14 | 0.00 | 0.12 | 0.00 | 0.00 | 27.0% |
| WM-82M | 1.24 | 0.00 | 0.24 | 0.01 | 0.24 | 0.00 | 0.00 | 49.9% |
| WM-81M | 1.89 | 0.00 | 0.32 | 0.03 | 0.44 | 0.00 | 0.00 | 82.2% |
| WM-80M | 2.14 | 0.00 | 0.38 | 0.06 | 0.55 | 0.00 | 0.00 | 101.4% |

The effect of Temperature upon the gas composition at various temperatures may be found in Figures 2 through 4. Each figure shows the effect of increasing temperature on the composition of the gas effluent, as well as the total carbon gasification.

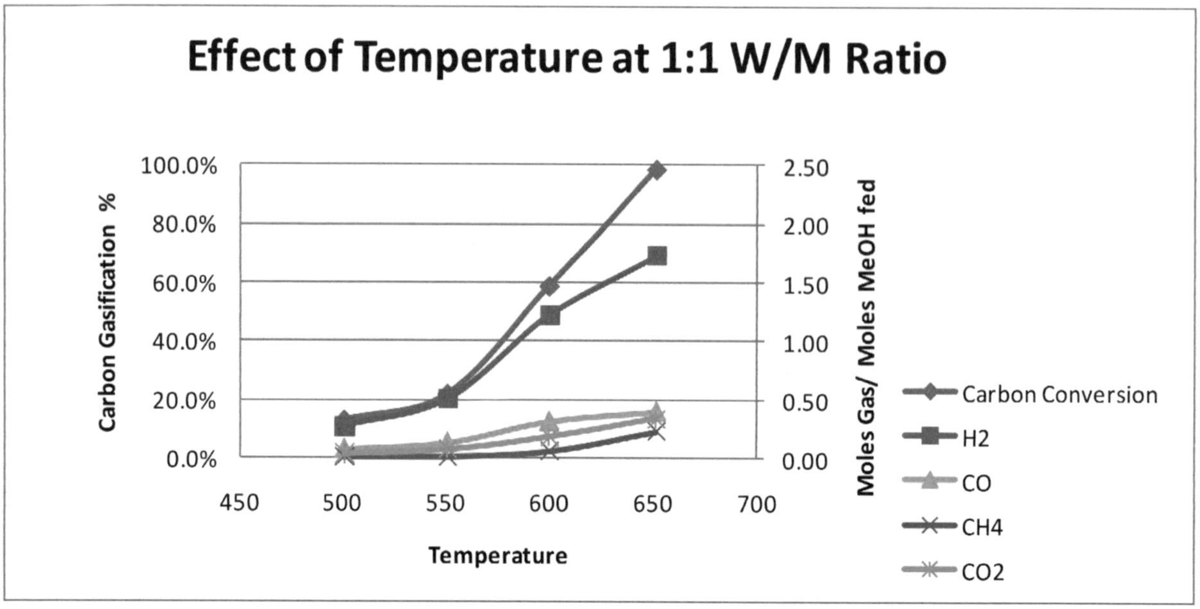


**Figure 2 Effect of Temperature at 1:1 W/M Ratio**

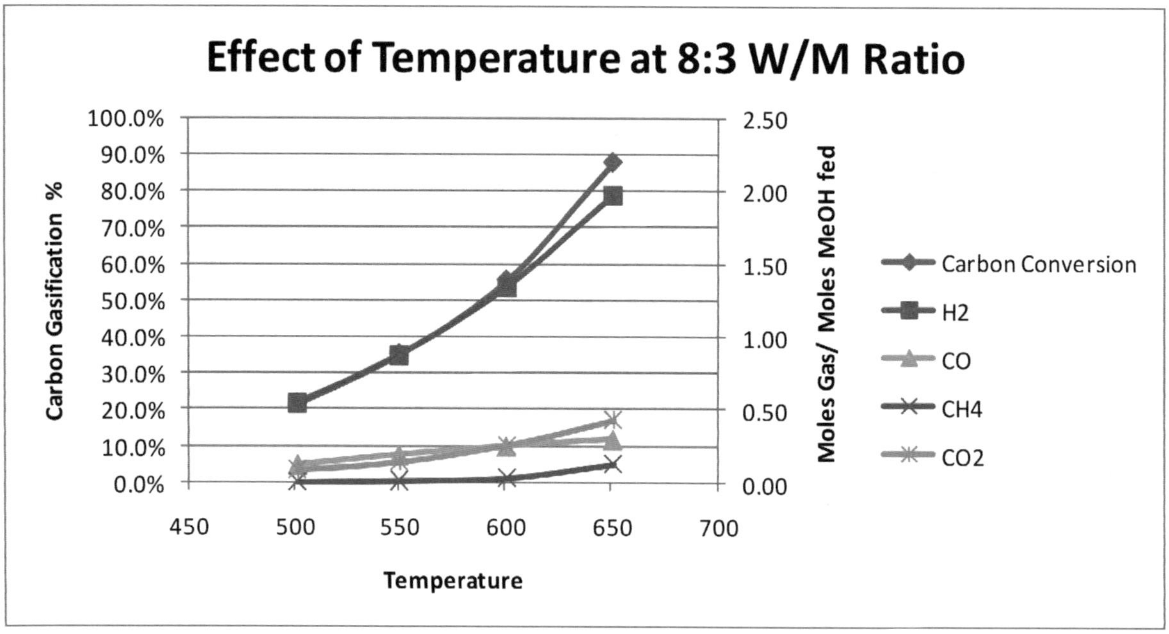


**Figure 3 Effect of Temperature at 8:3 W/M Ratio**

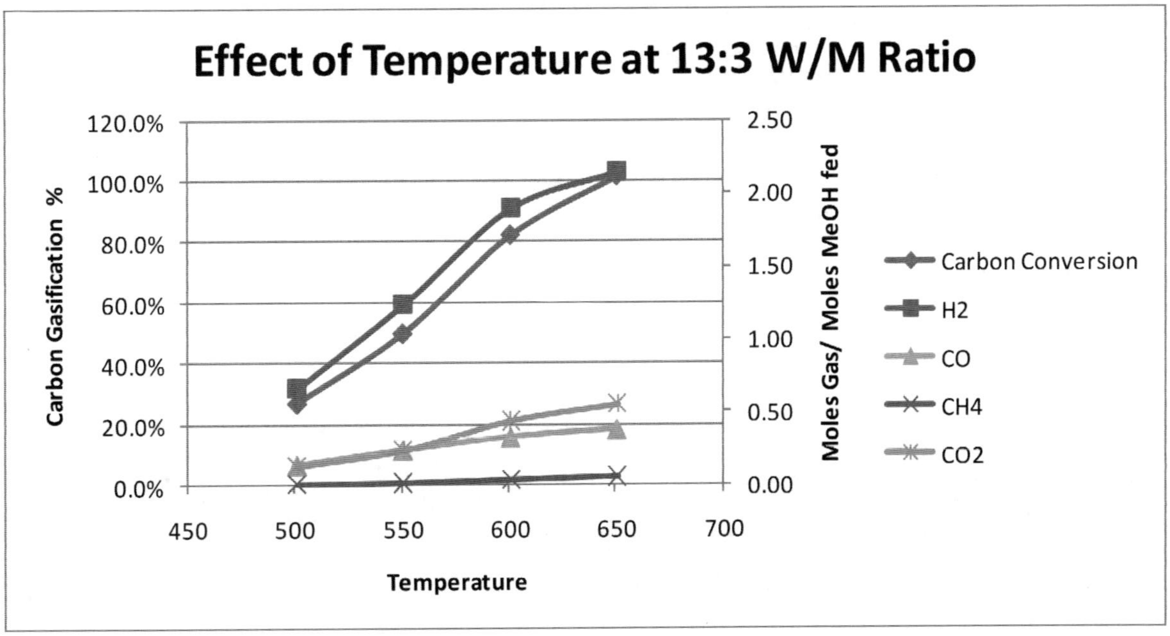


**Figure 4 Effect of Temperature at 13:3 W/M Ratio**

On all three graphs there is increase in total carbon gasification with increase in temperature, which makes sense with methanol decomposition reaction being an endothermic reaction. Also the amount of each type of gas produced per mole of methanol fed increases with temperature for all water-to-methanol ratios. This is due to the fact that methanol decomposition is a gateway reaction for all of the others since they all have carbon oxides as reactants. If not as much methanol decomposition happens, all gas numbers will decrease.

Trying to make conclusions about individual gas species is a little more difficult since the amount of gas produced per mole of methanol fed is increasing for all species. However, there does seem to be a trend with carbon monoxide increasing slower and carbon monoxide increasing faster at higher temperatures, which could be indicative of more forward WGS reaction occurring. There also seems to be a possible trend with hydrogen increasing slower at higher temperatures while methane increases faster, which would mean an increase in methanation. The trend is too to slight to say anything definitively however.

Figures 5 through 8 show the effects of changing water-to-methanol ratio at constant temperatures. Like the previous graphs, changes in carbon gasification and moles of gas per mole of methanol fed will be shown.

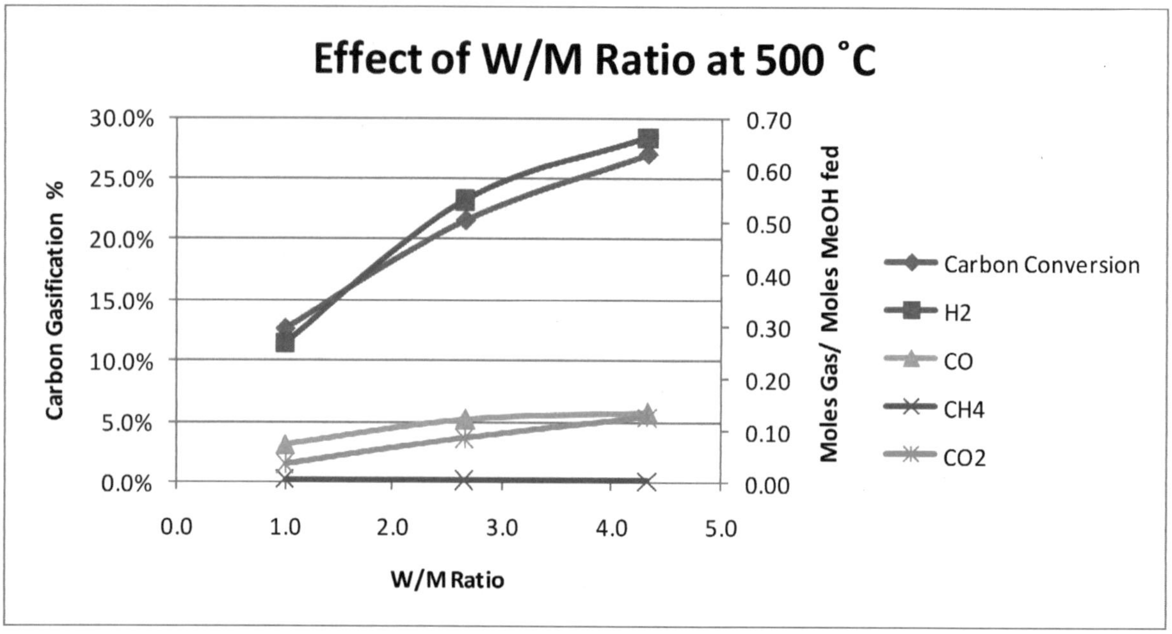


**Figure 5 The Effect of W/M Ration at 500 °C**

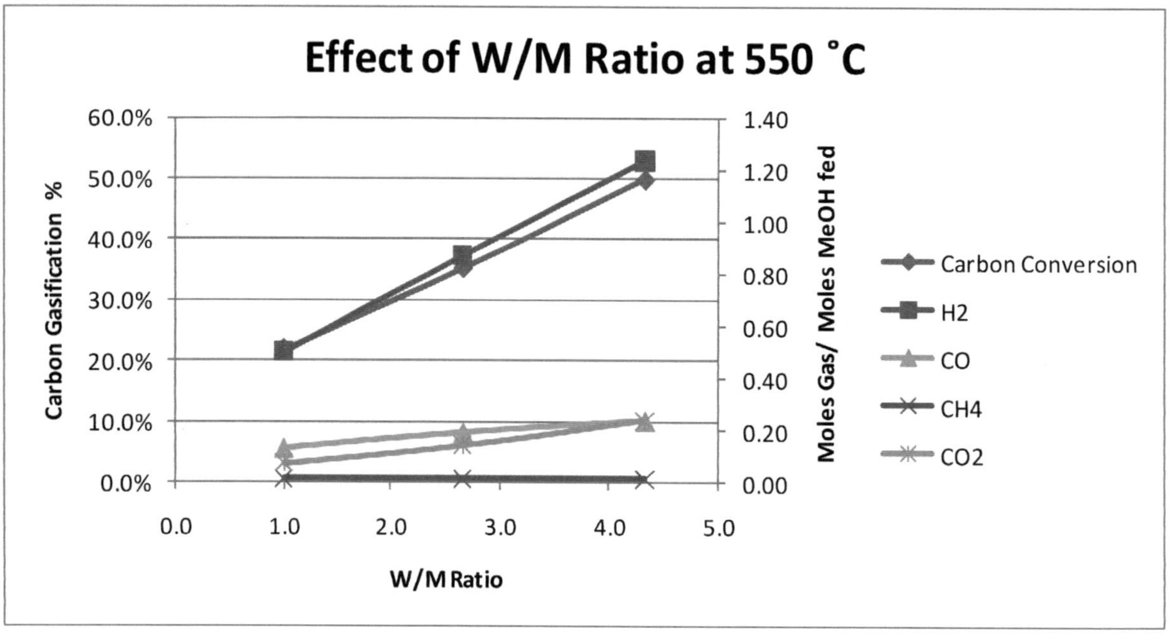


**Figure 6 The Effect of W/M Ration at 550 °C**

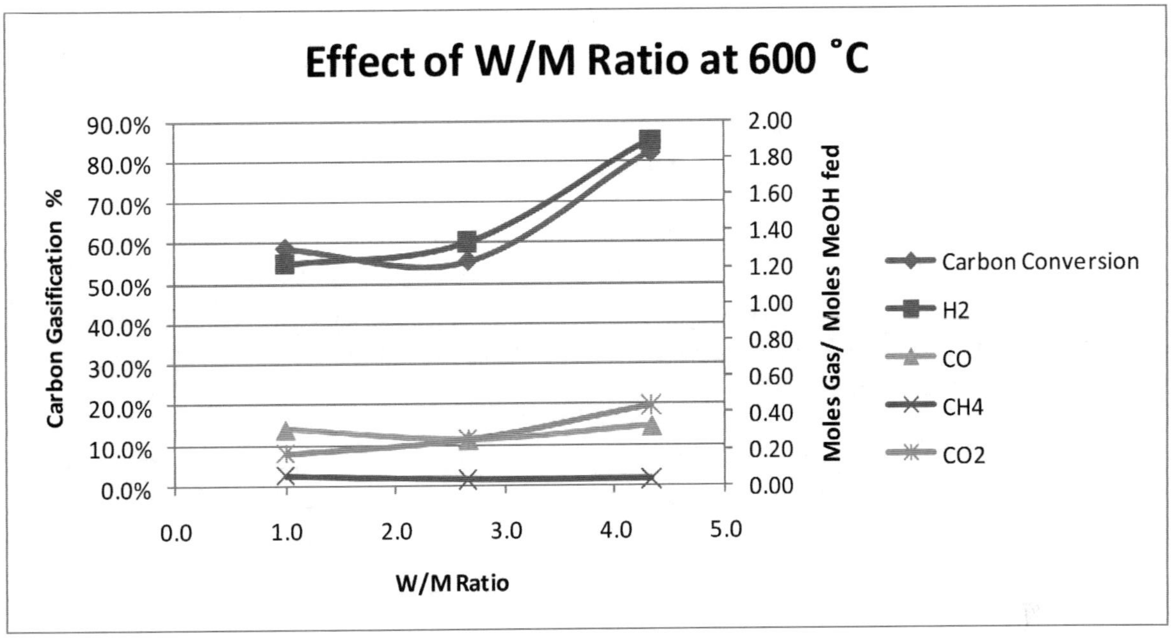


**Figure 7 The Effect of W/M Ration at 600 °C**

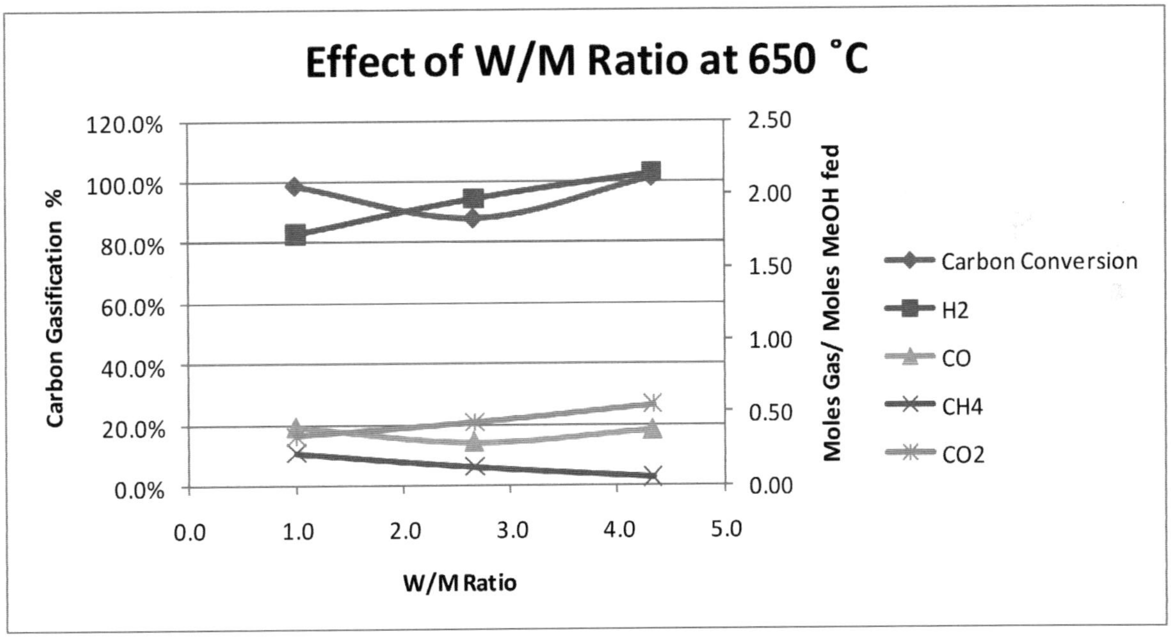


**Figure 8 The Effect of W/M Ration at 650 °C**

The effects of water-to-methanol (W/M) ratio on the reformation of methanol seem to vary a good amount with temperature. At 500 °C and 550 °C, carbon gasification and the gas produced per mole of methanol fed are all increasing as the solution becomes more dilute. Carbon dioxide seems to increase faster than carbon monoxide as the solution becomes more dilute, meaning more WGS reaction. There seems to be little to no methanol production at these temperatures, which could be attributed to the fact that overall there is little gasification, which means less methanol decomposition which is a precursor for methanation.

The effect of W/M ratio at temperatures of 600 ˚C and 650 ˚C is much more interesting. In both graphs there seems to be a drop in carbon gasification for the 8:3 W/M ratio solution. The drop is not too severe, so it could be experimental error, but since the drop occurs on both graphs it could be something else. More experiments should be performed at intermediate W/M ratios to see if there is in fact a trend there. As of results from the gas composition, methane steadily drops and other gases increase as the solution becomes more dilute. This would correspond to decrease in methanation, and an increase in the reformation of methanol.

## *Conclusion*

The results of these experiments show that with increasing temperature and more dilution of the water-methanol solution, more carbon gasification occurs and more of each gaseous species is produced. This is due to the increased amount of methanol decomposition occurring, which is needed to propagate the other reactions. An increase in forward WGS is shown with increasing temperature and dilution.

Future work that needs to be explored includes running experiments at intermediate W/M ratios at 600 ˚C and 650 ˚C to determine if the drop in carbon conversion was an error, or part of a trend. Also repeats of other experimental runs will be conducted to ensure repeatability. Finally crude glycerol experiments will be conducted, using these experiments in addition to similar pure glycerin experiments as guides for the behavior of the how crude glycerol mixture will react to supercritical water.

## *References*


(1) Johnson, D.T.; Taconi, K.A. The Glycerin Glut: Options for the Value-Added Conversion of Crude Glycerol Resulting from Biodiesel Production. Published online Oct. 1, 2007 in Wiley InerScience

(2) Nilles, Dave. Combating the Glycerin Glut. *Biodiesel Magazine*.[online] **2006**, *25* Article Archive http://www.biodieselmagazine.com/article.jsp?article_id=1123&q=&page=all (accessed September 15, 2009)

(3) Picou, J.W.; Wenzel, J.; Stever, M.; Bouquet, J.; Lee, S. Synthesis gas production by noncatalytic reforming of glycerin in supercritical water: Effect of temperature. Presented at the AIChE Fall National Meeting, Nashville, TN, November 2009.

(4) Boukis, N.; Diem V.; Habicht, W.; Dinjus, E. Methanol Reforming in Supercritical Water. *Ind. Eng. Chem. Res.* **2003,** *42,* 728-735

(5) Lee, S. *Alternative Fuels*, Taylor & Francis, Philadelphia, **1996**; pp.140-151.

(6) Methanol Pure Component Properties. Korean Thermophysical Properties Databank. http://www.cheric.org/research/kdb/hcprop/showprop.php?cmpid=817

(7) Water Phase Change Properties. National Institute of Standards and Technology. http://webbook.nist.gov/cgi/cbook.cgi?ID=C7732185&Units=SI&Mask=4#Thermo-Phase

(8) *Inconel® Alloy 625*, Special Metals. http://www.specialmetals.com/documents/Inconel%20alloy%20625.pdf accessed 9/12/10

# (574d) Kinetics and Characteristics of Steam Hydrogasification of Biomass by Using An Inverted Stirred Batch Reactor

**Xin Fan**[1], Megha Patel[2], Chan S. Park[2] and Joseph M. Norbeck[1], (1)Center for Environmental Research and Technology, College of Engineering, University of California, Riverside, Riverside, CA, (2)Center for Environmental Research and Technology, University of California, Riverside, Riverside, CA

Our research group is developing a multi-step thermal chemical process to convert carbonaceous matter into liquid fuels and chemicals. It is based on steam hydrogasification and is referred to in this study as the CE-CERT process. This process appears to offer several advantages in carbon efficiency and cost compared to other mainstream gasification technologies. The initial reactor in this process is the Steam Hydro-gasification Reactor for Steam Hydro-gasification Reaction (SHR) to produce the synthetic output gas. This study will describe the design, construction details and application of a new novel laboratory-scale batch reactor that allows for further fundamental laboratory studies of the kinetics and characteristics in the SHR process to complement other on-going pilot scale development of the CE-CERT process.

The newly designed batch reactor is an inverted stirred system with a pressure-driven feedstock loader that can rapidly load the feedstock material into the reactor at defined temperatures. This is compared to our previous batch reactor system that required long delays when the reactor is heated with the feedstock slowly reaching the desired reaction conditions. This unique design results in rapid heating of the feedstock to the target temperature and more closely represents the thermal environment of our continuously fed fluidized bed reactor being developed for the pilot and commercialized size systems. This allows for kinetic and product studies at a fraction of the time and cost. Additionally, the new batch reactor is built with a non-reactive, non-catalytic quartz vessel inserted into the metal reactor vessel to hopefully minimize any potential metal wall effects on the kinetics. Initial studies using the newly designed reactor shows an increase in several reaction kinetic rates; we believe the consequence of the stable reaction temperatures and the rapid heating rate.

This study will also present the results of the steam hydrogasification of biomass (pinewood) at temperatures ranging from 933K to 973K, under a H2 environment and a H2O to carbon mass ratio 2.5:1 using the inverted stirred batch reactor. The results indicate that the influence of the rapid heating rate not only increased the overall carbon conversion within the batch SHR system but also significantly affected the kinetic rates of all of the SHR reactions. In particular, an increase in the reaction temperature promoted carbon conversion of the biomass and enhanced the formation rate of CH4 and CO in the SHR process.

See more of this Session: Advances in Gasification Research I

# Experimental Studies on Gasification of Pine Wood Shavings in a Downdraft Biomass Gasifier

Amita Ummadisingu[1], Pratik Sheth[2] and B. V. Babu[3a],

[1]Student, Chemical Engineering Group

[2]Assistant Professor, Chemical Engineering Group

[3]Professor – Chemical Engineering Group, and Dean – Student Welfare Division & Faculty Division-1
Phone: +91-1596-24507 Ext 257; +91-1596-515257 Fax: +91-1596-244183;
Email: bvbabu@bits-pilani.ac.in, profbvbabu@gmail.com

Birla Institute of Technology and Science (BITS), PILANI – 333 031, India

**Abstract**

Biomass is the oldest form of energy used by human beings, mainly in the form of wood. It is a renewable source of energy and has many advantages from an ecological point of view. The direct combustion of biomass generates concern among environmentalists, as it is incomplete and inconsistent and may produce organic particulate matter, carbon monoxide and other organic gases. Biomass gasification has attracted the greatest interest as it offers a higher efficiency as compared to combustion and pyrolysis. Gasification is the process of conversion of solid carbonaceous fuel into combustible gas by partial combustion. Various groups of researchers have carried out experimental studies on downdraft biomass gasifier with different biomass materials such as wood, wood chips, hazelnut shell, agricultural wastes, saw dust, etc. It is reported that bridging is a normal occurrence in the gasifier. It occurs mainly in the pyrolysis zone of the downdraft biomass gasifier. In the present study, wood shavings of pine wood are used as biomass in a downdraft biomass gasifier. To overcome bridging, the design of the biomass gasifier is modified by incorporating a shaking mechanism. With this modified design of the biomass gasifier, experiments are carried out with pine wood (*Pinus roxburghii*) shavings as biomass covering wide range of operating conditions. The performance of the biomass gasifier system is evaluated in terms of equivalence ratio, producer gas composition, calorific value of the producer gas, gas production rate, zone temperatures and cold gas efficiency.

**Keywords**: *Biomass Gasification, Producer Gas, Pinus roxburghii, Equivalence ratio.*

## 1. Introduction

There are many ways by which the energy, available in abundance around us, can be stored, converted and amplified for our use. Energy sources will play an important role in the world's future. The cost of producing energy from biomass fuels is less than that from fossil fuels. The term biomass is used to describe all the biologically produced matter. Biomass is the oldest form of energy used by human beings, mainly in the form of wood. It has either been burned directly in furnaces, or processed to increase its energy content (Haykiri-Açma, 2003). Biomass is useful to meet different kinds of energy needs, including fueling vehicles, providing process heat for

[a] Corresponding Author

industrial facilities, generating electricity and heating homes. It is a renewable source of energy and has many advantages from an ecological point of view (Babu and Sheth, 2006). Biomass gasification is the incomplete combustion of biomass resulting in production of combustible gases consisting of carbon monoxide, hydrogen, methane, carbon dioxide and nitrogen (Sheth and Babu, 2009a). This mixture is called producer gas. It is burnt to produce process heat and steam or used in gas turbines to produce electricity (Sheth and Babu, 2006; Sheth and Babu, 2009b). The concentration of hydrogen in producer gas depends upon many factors such as moisture content of biomass, type and composition of biomass, operating conditions, configuration of the biomass gasifier, etc. (Sheth and Babu, 2010).

Dogru et al. (2002) carried out gasification studies using hazelnut shell as a biomass. Jayah et al. (2003) used a downdraft biomass gasifier using rubber wood as biomass in a tea drying unit. In order to assess the potential of the saw dust as a biomass material, a fixed bed, downdraft, stratified and open top gasifier was used by Wander et al. (2004). Zainal et al. (2002) performed experimental studies on a downdraft biomass gasifier using wood chips and charcoal. Sheth and Babu (2009b) carried out experiments with waste wood blocks of *Dalbergia sisso* as biomass in a downdraft biomass gasifier to generate the producer gas. In the present study, the composition of the producer gas is measured using a gas chromatograph and the effects of equivalence ratio on producer gas composition, calorific value of the producer gas, gas production rate, zone temperatures and cold gas efficiency are studied. It is reported that bridging of biomass loaded in the gasifier is a common problem in the operation of the gasifier. This problem is mainly found with low density biomass materials such as rice husk, wood shavings, etc. It occurs mainly in the pyrolysis zone of the downdraft biomass gasifier.

In the present study, gasification experiments with wood shavings of *Pinus roxburghii*, commonly called Chir wood, as biomass are carried out in downdraft biomass gasifier. The wood shavings of the pine wood, a waste material, are collected from the carpentry section of the institute's workshop. The problem of bridging was faced in the initial trial runs. The biomass gasifier is modified by incorporating a shaking mechanism to reduce bridging of biomass in the pyrolysis zone of the biomass gasifier. The experiments are carried out on the modified gasifier with *Pinus roxburghii* wood shavings as biomass, covering a wide range of operating conditions. The effects of varying parameters such as air flow rate and moisture content on the quality of the producer gas generated are studied. The performance of the biomass gasifier system is evaluated in terms of equivalence ratio, producer gas composition, calorific value of the producer gas, gas production rate, zone temperatures and cold gas efficiency.

## 2. Biomass Gasification

Four distinct processes take place in a gasifier as the fuel makes its way to gasification. They are drying of fuel, pyrolysis, combustion and reduction (Babu and Sheth, 2006b). An Imbert downdraft biomass gasifier is used in the present study. It has throated combustion zone and different area of cross section for pyrolysis and reduction zone unlike a stratified downdraft biomass gasifier in which gasifier cross sectional area is uniform throughout the gasifier (Sheth and Babu, 2009b). The products of wood pyrolysis depend upon temperature, pressure, residence time and heat losses. Pyrolysis produces large quantities of tar and gases containing carbon dioxide. In downdraft gasifiers, pyrolysed gas generated in pyrolysis zone gets mixed with the moisture coming from drying zone and flows downward. The tars have to go through combustion and reduction zone and are partially broken down. The combustible substance of a solid fuel is usually composed of elements carbon, hydrogen and oxygen. In combustion, carbon dioxide is obtained from carbon in fuel and water is obtained from the hydrogen, usually as steam. The combustion reaction is exothermic. The heat generated in the combustion zone is transferred to the pyrolysis and drying zone (Sheth and Babu, 2009b). The products of partial combustion (water, carbon dioxide and uncombusted partially cracked pyrolysis products) pass through a red-hot charcoal bed where endothermic reduction reactions take place (Rajvanshi, 1986). Biomass is fed to the gasifier and oxidized in the zone where air is continuously supplied from two air nozzles. The dimensional details of the biomass gasifier used in the present study and experimental procedure are reported in Sheth and Babu (2009b).

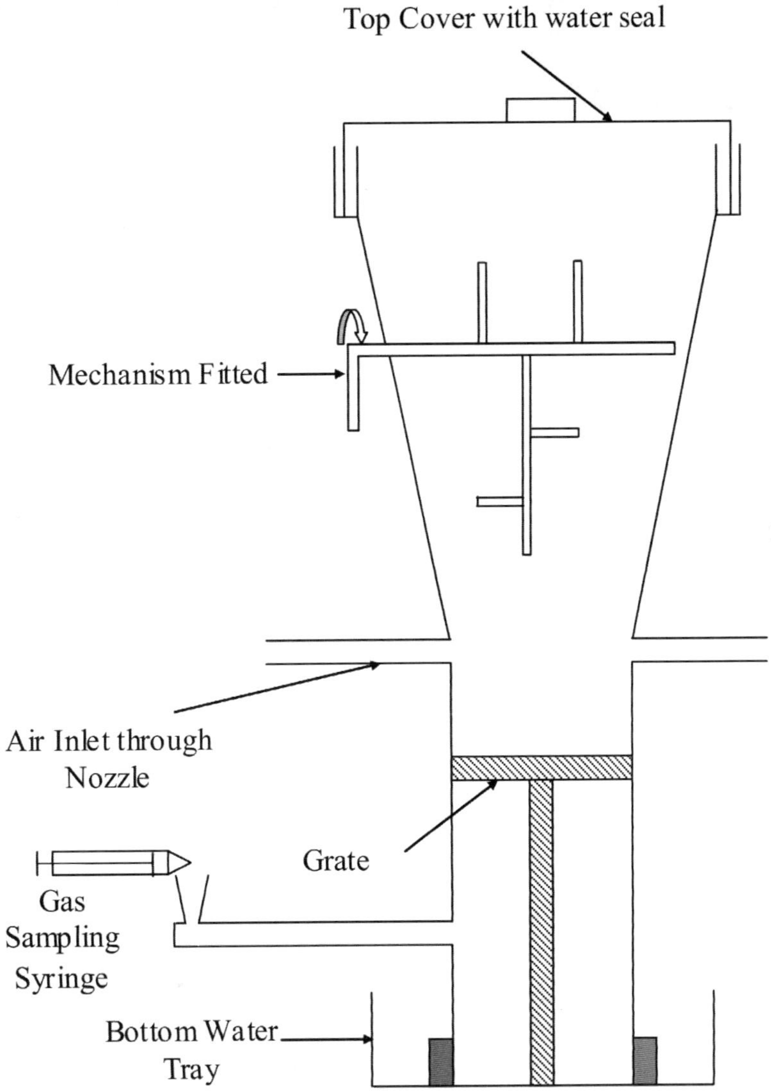


**Figure 1: Schematic of modified downdraft gasifier**

As soon as the biomass in combustion zone gets consumed, biomass in the pyrolysis zone flows downward due to gravity. If the biomass doesn't flow downward easily, it leads to bridging.

There are two main reasons for bridging *viz.* (1) the bulk density of the biomass is too low, and (2) cone angle is less than angle of repose of the biomass. The second criterion is generally taken care of while designing the downdraft biomass gasifier. However, while using light weight biomasses such as wood shavings, moist saw dust, rice husk, etc. in a downdraft biomass gasifier, the practical operating problem of bridging is faced. During the initial experimental runs, bridging occurs and gasifier cannot be operated continuously for more than 10 minutes. It is due to the low bulk density of the wood shavings of the pine wood. To overcome the bridging, the design of the biomass gasifier is modified by incorporating the shaking mechanism. The 'L' shape steel rod is placed in the gasifier, one end of which is extended along the axis of the gasifier up to the oxidation zone. Some small rods (in the radial directions up to the wall of the gasifier) are welded on its axial part of 'L' shaped rod at different heights. The other part of the rod is positioned to protrude through the side wall to the outside of the gasifier. Mechanical fitting is carried out to ensure that there is negligible leakage and free movement while the biomass of pyrolysis zone is shaken from outside the gasifier.

## 3. Results and Discussion

### 3.1. Producer gas composition and Temperature profiles

The physical properties of the *Pinus roxburghii* are listed in Table-1 (Jain, 1992).

**Table -1 Characteristics of Pinus roxburghii**

| **Physical Properties** | |
|---|---|
| Density (kg/m$^3$) | 480 |
| **Proximate Analysis (% by wt. dry basis)** | |
| Fixed Carbon (FC) | 54.82 |
| Volatile Matter (VM) | 43.26 |
| Ash | 1.91 |
| HHV (MJ/kg) | 23.17 |

The producer gas composition and the temperature profile in the gasifier for Experimental Run 1 are presented in sections 3.1.1 and 3.1.2. The analysis for the other experimental runs conducted is carried out in a similar manner.

*3.1.1. Producer gas composition*

The composition of the producer gas, sampled at five minute intervals during gasification experiments, is found using a gas chromatograph (NUCON 5765) with a thermal conductivity detector. The molar compositions of the producer gas obtained in Experimental Run 1 are shown in Fig.2. It is found that the quantities of CO and $H_2$ gases obtained are good during steady operation between 10 and 20 minutes. A large molar fraction of $N_2$ gas is observed at 5 minutes corresponding to startup and at 25 minutes from the start of the gasifier. The operation of gasifier is stopped at 25 minutes for all the experimental runs carried out in the present study.

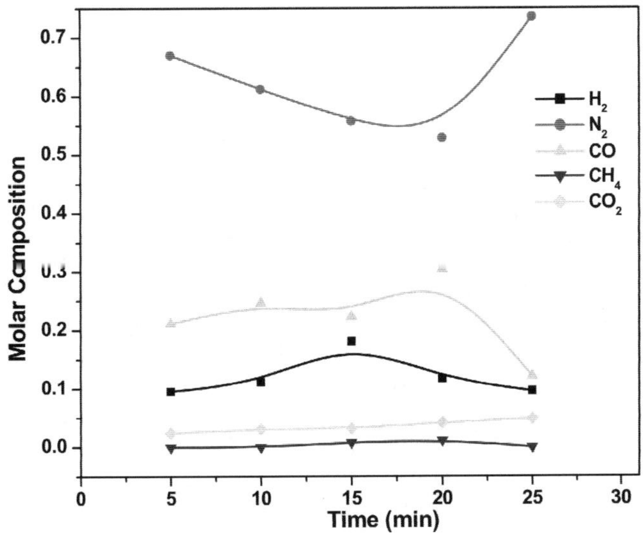


**Figure 2: Variation of molar composition with time for Experimental Run 1**

*3.1.2. Temperature Profile*

The temperatures displayed by various thermocouples embedded in the gasifier are noted during the experimental runs at time intervals of 5 minutes. Fig.3 shows the plot of these temperatures as a function of time. In the gasifier, the reduction zone is present between 0 – 90 mm from the grate, while the oxidation zone lies between 90 -150 mm from the grate. Between 200 – 450 mm from the grate, the pyrolysis and drying zones are encountered. Fig.4 shows the temperature at various heights from the grate at the centre ($r$ = 0) of the gasifier and at half the radius ($r = R/2$) of the gasifier. The temperature of the oxidation zone (distance from grate (dfg) = 130 at $r$ = 0) varies from 750 – 1150°C, while that of the pyrolysis zone (dfg = 245 at $r$ =0) varies from 230 – 550°C. For the drying zone (dfg = 430 at $r$ = 0), temperature varies in the range of 100 – 300°C. It can be seen that the temperatures at $r$ = 0 are generally greater than those at $r = R/2$ during steady operation of the gasifier.

The mechanism for reduction of bridging, described in section 2 of this paper, is utilized every five minutes during the operation. This moves part of the unconsumed biomass into the lower regions of the gasifier hence allowing continuation of operation of the gasifier. It can be seen in Fig.3 that the temperature of the oxidation zone falls from around 1060°C to around 700-800°C in about 20 minutes. The temperatures of the upper regions of the gasifier, such as the pyrolysis zone and the drying zone increase. This occurs when the biomass in the oxidation zone is consumed and some bridging in the uppermost regions of the gasifier, beyond the reach of the mechanism, prevents the unconsumed biomass from entering the oxidation zone. Fig.4 shows the temperature variation as a function of distance from grate at time intervals of 5 minutes for Experimental Run 1. It provides an idea of the temperatures at different points covering the height of the gasifier at intervals of 5 minutes.

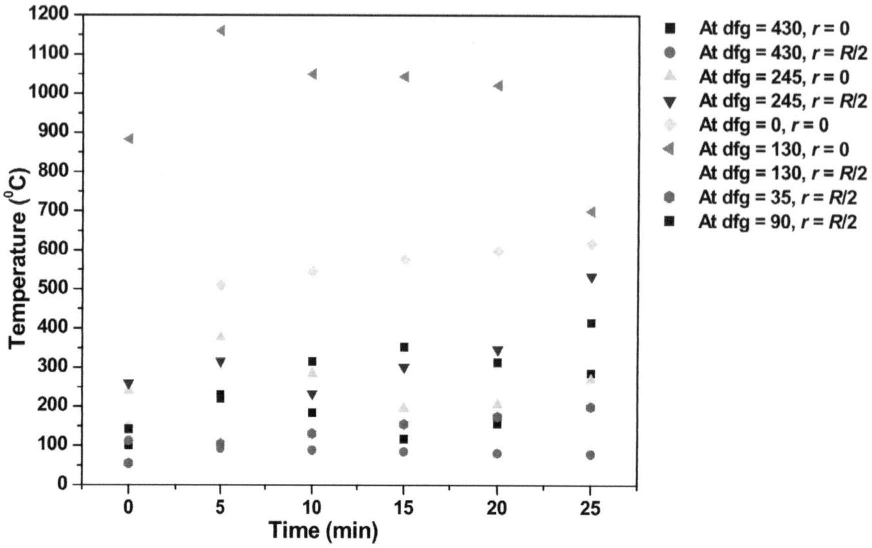


**Figure 3: Temperature variation as a function of time for Experimental Run 1**

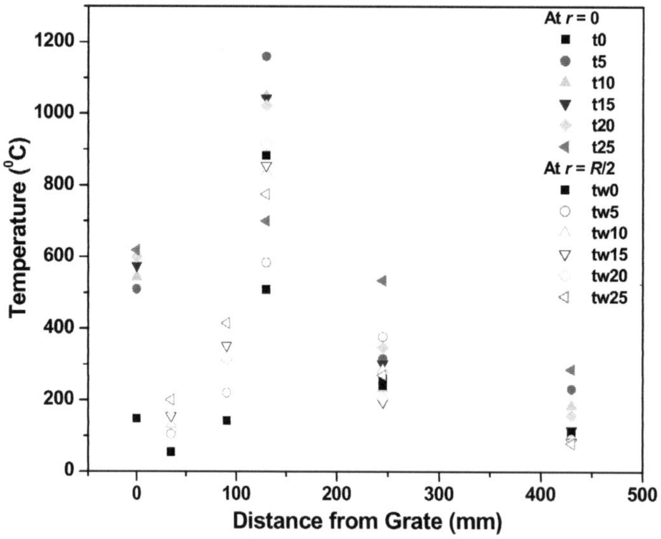


**Figure 4: Temperature variation as a function of distance from grate at time intervals of 5 minutes for Experimental Run 1**

### 3.2. Analysis with respect to equivalence ratio

The details of parameters varied during present experimental study are given in Table-2. Biomass consumption rate is found to vary from 0.72 to 1.92 kg/h for an air flow rate varying from 1.54 to 2.16 $m^3$/h. Moisture content is varied from 0.0366 to 0.363 wt. fraction on wet basis. In order to reduce the number of parameters on which the performance of the biomass gasifier depends, an equivalence ratio is defined to reflect the combined effect of air flow rate,

rate of wood supply and duration of the run. The equivalence ratio ($\phi$) for each run is calculated by Eq. (1) (Reed and Das, 1988).

$$\text{Equivalence ratio } (\phi) = \frac{(\text{Air flow rate/ Biomass consumption rate})}{(\text{Air flow rate/ Biomass consumption rate})|_{\text{Stoichiometric}}} \quad (1)$$

The stoichiometric ratio of air flow rate to biomass consumption rate is 5.22 $m^3$ air/kg of wood (Zainal et al., 2002). The values of equivalence ratio are calculated for each run and reported in Table 2.

**Table 2: Details of the experimental runs for biomass gasification**

| Experimental Run No. | Air flow rate ($m^3$/h) | Initial moisture content (wt fraction, wet basis) | Biomass consumption Rate (kg/h) | Equivalence ratio($\Phi$) |
|---|---|---|---|---|
| 1 | 2.16 | 0.0366 | 1.92 | 0.2155 |
| 2 | 1.851 | 0.0366 | 1.7 | 0.1738 |
| 3 | 1.5425 | 0.0366 | 0.72 | 0.4104 |
| 4 | 1.5425 | 0.3490 | 1.44 | 0.2052 |
| 5 | 1.5425 | 0.3630 | 1.2 | 0.2462 |
| 6 | 1.5425 | 0.1608 | 0.9 | 0.3283 |

The performance of the biomass gasifier system is evaluated in terms of the producer gas composition, the calorific value of producer gas, zone temperatures, gas generation rate, and cold gas efficiency.

*3.2.1. Variation of producer gas composition with equivalence ratio*

The average gas composition for each experimental run is plotted against the equivalence ratio of the corresponding experimental run and shown in Fig.5. The average should take care of the variation with respect to time. It is noted that the molar fractions of nitrogen and carbon dioxide decrease with an increase in $\phi$ up to a value of $\phi = 0.21$. Molar fractions of nitrogen and carbon dioxide increase for higher values of $\phi$. The fractions of carbon monoxide and hydrogen increase up to $\phi = 0.21$ and then decrease while the opposite trend is observed for nitrogen and carbon dioxide. A higher value of $\phi$ indicates a higher air flow rate for a specific biomass consumption rate. This leads to a larger $CO_2$ production in combustion zone and a larger quantity of $N_2$ entering the gasifier through the air supply. The rate of reactions occurring in the reduction zone and length of the reduction zone control the conversion of $CO_2$ to CO. A greater amount of $CO_2$ is converted to CO and $H_2$ in the combustion zone corresponding to an increase in $\phi$ from 0.174 to 0.21. Thus, the fractions of CO and $H_2$ increase with $\phi$ till a value of $\phi = 0.21$ and fractions of $CO_2$ and $N_2$ decrease. The increase in $CO_2$ and decrease in CO & $H_2$ fractions for the equivalence ratio higher than 0.21 indicate that production of $CO_2$ in combustion zone exceeds the conversion capacity of reduction bed to convert it to CO. As mentioned earlier, for larger values of $\phi$ ranging from 0.21 to 0.41, a larger quantity of $N_2$ enters the gasifier through the air supply and hence an increase in $N_2$ fraction is observed.

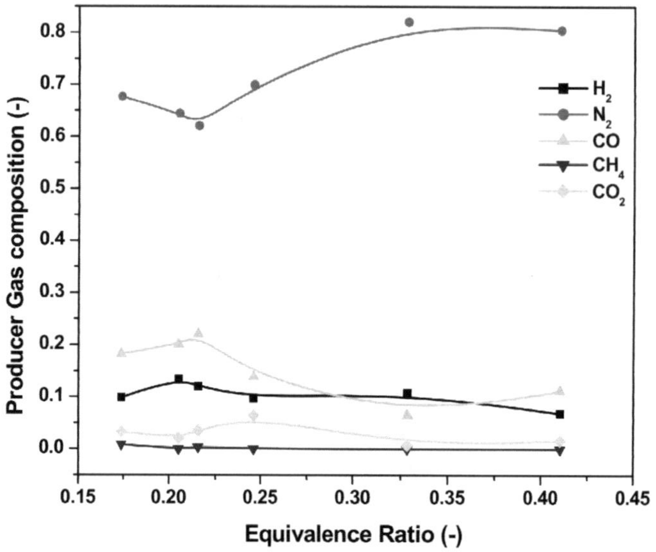


**Figure 5: Effect of equivalence ratio on producer gas composition.**

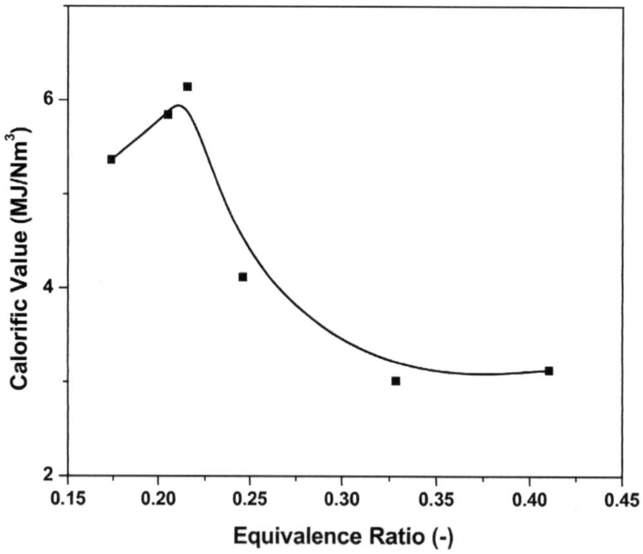


**Figure 6: Effect of equivalence ratio on calorific value of producer gas**

*3.2.2. Variation of calorific value with equivalence ratio*

The calorific value can be calculated based on the composition of the producer gas. As mentioned earlier, carbon monoxide and hydrogen are the main components of the producer gas. A greater concentration of carbon monoxide and hydrogen corresponds to a higher calorific value. The presence of methane in the gas increases the calorific value but as seen in Fig.5, the fraction of methane is very small. Variation of calorific value of the producer gas with equivalence ratio is presented in Fig.6. It is found that at an equivalence ratio of 0.174 the calorific value is the least at 5.36 MJ/Nm$^3$. With a small increase in the equivalence ratio at $\phi = 0.21$, the calorific value reaches a maximum of 6.14 MJ/Nm$^3$ and then follows a decreasing trend. Calorific value steadily decreases with a further increase in the equivalence ratio beyond 0.21. This is in agreement with the decreasing trends of carbon monoxide and hydrogen fractions with increase in $\phi$ value beyond 0.21 as discussed in section 3.2.1.

*3.2.3. Variation of zone temperatures with equivalence ratio*

The variation of the average temperature of pyrolysis and oxidation zone of the downdraft biomass gasifier with the equivalence ratio is shown in Fig.7. The temperature of oxidation zone is found to vary from 900°C to 1100°C and that of the pyrolysis zone between 150°C and 400°C. The oxidation zone temperature depends upon the heat released due to the biomass combustion and air flow rate. An increase in air flow rate increases the oxygen available for oxidation but also increases the amount of $N_2$ entering the gasifier. $N_2$ being inert, acts as a heat carrier and reduces the temperatures in the oxidation and pyrolysis zones. The maximum value of temperatures in the oxidation zone represents the optimum amount of equivalence ratio ($\phi = 0.21$). This is in agreement with the optimum equivalence ratio obtained in the previous sections based on producer gas composition and calorific value analysis.

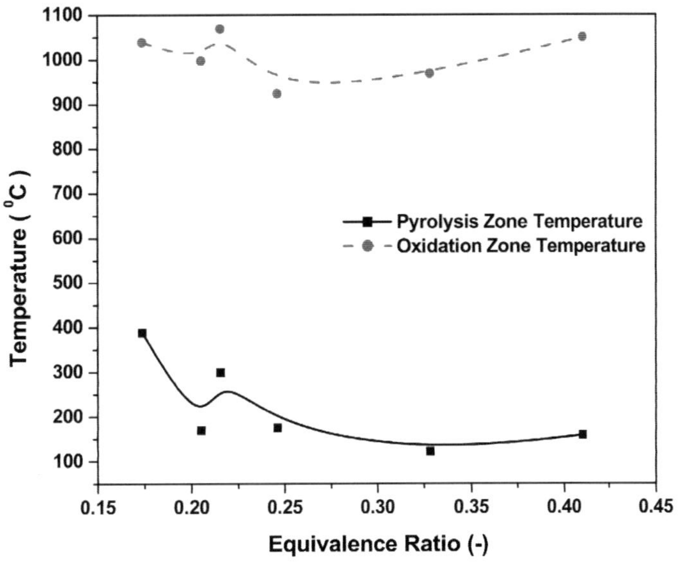


**Figure 7: Effect of equivalence ratio on zone temperatures.**

*3.2.4. Variation of gas production rate with equivalence ratio*

In section 3.2.1, we have seen that a higher equivalence ratio corresponds to a higher air flow rate for a specific biomass consumption rate. Fig.8 shows that with an increase in the equivalence ratio, producer gas production rate increases with maxima at an equivalence ratio of 0.21. After a certain value of the equivalence ratio, calorific value of the producer gas deteriorates due to higher amounts of carbon dioxide but the production rate of producer gas continues to increase. This increase in production rate of producer gas is due to the larger quantity of $N_2$ entering the gasifier through the air supply for larger equivalence ratios.

*3.2.5. Variation of cold gas efficiency with equivalence ratio*

Cold gas efficiency is defined as the ratio of energy of the producer gas per kg of biomass to the HHV of the biomass material. The cold gas efficiency depends upon the calorific value and the amount of producer gas released at constant HHV of biomass (Sheth and Babu, 2009b). Fig.8 shows that the amount of producer gas increases continuously and the cold gas efficiency passes through maxima at $\phi = 0.21$ with an increase in the equivalence ratio.

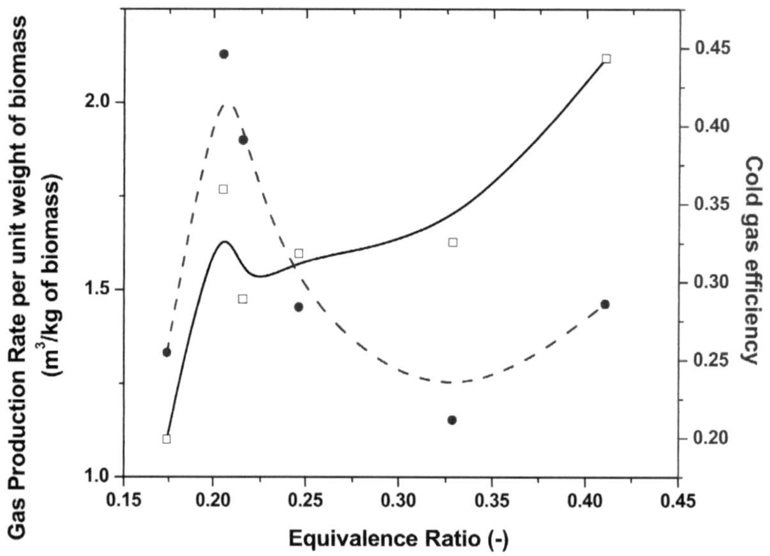


**Figure 8: Effect of equivalence ratio on producer gas production rate per unit weight of biomass and on cold gas efficiency.**

## 4. Conclusions

- Good amounts of CO and $H_2$ gases are obtained during steady operation between 10 and 20 minutes. The percentage of $H_2$ gas obtained is between 6.8% and 13.4% and that of CO is 6.5 % to 22%.
- Molar fractions of $N_2$ and $CO_2$ decrease up to an equivalence ratio of 0.21. For larger values of $\phi$ ranging from 0.21 to 0.41, there is an increase in $N_2$ and $CO_2$ fractions. This is opposite to the trend observed for $H_2$ and CO molar fractions.
- The temperature of oxidation zone varies from 900°C to 1150°C and that of pyrolysis zone varies between 150°C and 400°C. At an equivalence ratio of 0.21, the oxidation zone temperature passes through maxima.
- The trend of calorific value, producer gas production rate and cold gas efficiency variation is similar to that of variation of CO and $H_2$ fractions with an increase in equivalence ratio. With an increase in the equivalence ratio, calorific value, producer gas production rate and the calorific value have maxima at an equivalence ratio of 0.21.
- The optimal equivalence ratio for *Pinus roxburghii* wood shavings in the downdraft gasifier studied is 0.21.

## 5. References


Babu, B.V., Chaurasia, A.S., 2004. Heat Transfer and Kinetics in the Pyrolysis of Shrinking Biomass Particle. *Chemical Engineering Science*, **59**, 1999-2012.

Babu, B.V., Sheth, P.N., 2006. Modeling and Simulation of Reduction Zone of Downdraft Biomass Gasifier: Effect of Char Reactivity Factor. *Energy Conversion and Management*, **47**, 2602-2611.

Babu, B.V.; Sheth, P.N., 2009. Experimental Studies on Downdraft Biomass Gasifier. *Proceedings of International Symposium & 62nd Annual Session of IIChE in association with International Partners (CHEMCON-2009)*, Andhra University, Visakhapatnam, December 27-30, 2009.

Dogru, M., Howrath, C.R., Akay, G., Keskinler, B., Malik, A.A., 2002. Gasification of hazelnut shells in a downdraft gasifier. *Energy*, **27**, 415-427.

Haykiri-Açma, H., 2003. Combustion characteristics of different biomass materials. *Energy Conversion and Management*, **44**, 155-162.

Jain, R. K., 1992. Fuelwood Characteristics of Certain Hardwood and Softwood Tree Species of India. *Bioresource Technology,* **41**, 129-133.

Jayah, T.H., Aye, Lu, Fuller, R. J., Stewart, D.F., 2003. Computer simulation of a downdraft wood gasifier for tea drying. *Biomass and Bio energy*, **25**, 459-469.

Koufopanos, C. A., Maschio, G. and Lucchesi, A., 1989. Kinetic modeling of the Pyrolysis of Biomass and Biomass Components. *The Canadian Journal of Chemical Engineering*, **67**, 75-83.

Rajvanshi, A.K., 1986. Biomass Gasification, Alternative Energy in Agriculture, Vol:II, Ed. D. Yogi Goswami, CRC Press, 83-102, Chapter 4.

Reed, T.B., Das, A., 1988. Handbook of Biomass Downdraft Gasifier Engine Systems, The Biomass Energy Foundation Press, Colorado.

Sheth, P.N., Babu, B.V., 2006. Kinetic Modeling of the Pyrolysis of Biomass. *Proceedings of National Conference on Environmental Conservation* (NCEC-2006), BITS-Pilani, September 1-3, 2006, 453-458.

Sheth, P.N., Babu, B.V., 2009a. Differential Evolution Approach for obtaining Kinetic Parameters in Non isothermal Pyrolysis of Biomass. *Materials and Manufacturing Processes*, **24**, 1-6.

Sheth, P.N., Babu, B.V., 2009b. Experimental studies on producer gas generation from wood waste in a downdraft biomass Gasifier. *Bioresource Technology* , **100**, 3127-3133.

Sheth P.N., Babu, B.V., 2010. Production of hydrogen energy through biomass (waste wood) gasification. *International Journal of Hydrogen Energy*, Article in Press. doi:10.1016/j.ijhydene.2010.03.009 .

Wander, P.R., Altafini, C.R., Barreto, R.M., 2004. Assessment of a small saw dust gasification unit. *Biomass and Bioenergy*, **27**, 467-476.

Zainal, Z.A., Ali, R., Quadir, G., Seetharamu, K.N., 2002. Experimental Investigations of a downdraft biomass gasifier. *Biomass and Bioenergy*, **23**, 283-289.

# (574f) Sulfuric Compounds From Biomass in a Fluidized Bed Gasifier

**Brett Christensen**[1], Kevin Whitty[2] and Dan Sweeney[2], (1)Institute for Clean & Secure Energy, The University of Utah, Salt Lake City, UT, (2)Institute for Clean and Secure Energy, The University of Utah, Salt Lake City, UT

Biomass gasification is an attractive focus of research for several reasons, including: low pollution, feedstock availability and liquid fuel production. As the world turns its focus to cleaner and more efficient forms of energy, gasification is a viable option that can be implemented in various regions without drastic effects. One area that demands further research is the catalytic poisons created during the gasification process that effects subsequent synthesis gas cleaning to remove tars. Sulfuric compounds greatly inhibit a catalysts ability to crack tars during the cleaning process. Though biomass contains a relatively small amount of sulfur, its effects are a large obstacle in removing tars and producing liquid fuels. The most common sulfur containing compounds are hydrogen sulfide, carbonyl sulfide, thiophene and sulfur dioxide. Analysis of sulfur compounds at varying temperature, pressure and stoichiometry within a fluidized bed gasifier are explained within.

See more of this Session: Advances in Gasification Research I
See more of this Group/Topical: Topical E: High Temperature Environmentally Sustainable Energy Processes (sessions joint with the Environmental Division)

# (624a) Development of Coal Devolatilization and Secondary Pyrolysis Kinetics Models for CFD Codes

**Kiran Chaudhari**[1], Richard Turton[1], Chris Guenther[2] and Ronald Breault[3], (1)Chemical Engineering, West Virginia University, Morgantown, WV, (2)National Energy Technology Laboratory, U.S. Department of Energy, Morgantown, WV, (3)Gasification and Combustion Products Division, National Energy Technology Laboratory, Morgantown, WV

The objective of this research is to implement detailed kinetic expressions for the gasification of a wide variety of coals in existing CFD codes such as Multiphase Flow with Interphase Exchange (MFIX). These expressions describe the fundamental steps taking place in the gasification of coal, namely, coal devolatilization, tar-gas chemistry, and soot formation. For this purpose, the data generated by PC Coal Lab (licensed by Niksa Energy Associates, LLC) will be used to simulate the gasification of various coals in CFD codes. The CFD code of most interest is MFIX, which is used to describe the hydrodynamics, heat transfer and chemical reactions in the reacting fluid-solids systems comprising the contents of a gasifier. The implementation of coal gasification kinetics in MFIX is currently done through a set of subroutines making up the module Carbonaceous Chemistry for Continuum Modeling ($C_3M$) code. C $_3$M has default gasification kinetics for only a few coals. While on the other hand, PC Coal lab can predict the gasification kinetics for over 2000 coal species. Currently, there is no soot formation mechanism available in MFIX, hence the following mechanism for soot formation was introduced: Tar ↔PAH, Tar→ Soot, PAH →Soot. This project focuses on the development of a seamless connection between PC Coal Lab and the $C_3M$ .The interface will be designed to allow MFIX to transfer information through $C_3M$ to PC Coal Lab, run PC Coal Lab with the input data from MFIX, send kinetic information back to $C_3M$ in a form that allows $C_3M$ to update and continue the MFIX simulation using updated parameters from PC Coal Lab. Current work is focused on developing similar expressions for biomass and these results will also be discussed.

See more of this Session: Advances in Gasification Research II
See more of this Group/Topical: Topical E: High Temperature Environmentally Sustainable Energy Processes (sessions joint with the Environmental Division)

# (624b) Application of the Direct Quadrature Method of Moments to Large Eddy Simulations of Coal Gasification

**Charles Reid**[1], Jeremy N. Thornock[2] and Philip J. Smith[1], (1)Chemical Engineering, University of Utah, Salt Lake City, UT, (2)University of Utah, Salt Lake City, UT

Utilization of domestic sources of fuel such as coal is growing increasingly important, but under increasing scrutiny. There is a critical need to retrofit existing industrial and applied-scale facilities using coal to meet regulatory requirements, and to explore new technologies and techniques for carbon emission reduction. With the continually increasing power of computational resources, predictive simulation tools are playing an increasingly important role in this process. To this end, an Eulerian dispersed-phase model, the direct quadrature method of moments (DQMOM), has been implemented in a massively-parallel large eddy simulation (LES) code, and applied to coal gasification. The DQMOM formulation for gasification of coal particles must characterize the multivariate number density function (NDF) as a function of several particle independent variables (internal coordinates), including particle composition, particle temperature, and particle size. Next, several issues relevant to the implementation of DQMOM are addressed. Preliminary LES gasification results are presented, compared to experimental data, and discussed. A successful implementation of DQMOM in the context of massively parallel large-eddy simulations of coal gasification is demonstrated. Several recent developments in DQMOM theory help pave the way for application to complex coupled physics problems such as coal gasification.

See more of this Session: Advances in Gasification Research II
See more of this Group/Topical: Topical E: High Temperature Environmentally Sustainable Energy Processes (sessions joint with the Environmental Division)

# CFD SIMULATION OF COAL GASIFICATION IN AN ENTRAINED-FLOW GASIFIER

*Vikram Sreedharan, Aalborg University, Esbjerg, Denmark*
*Dr. Bjørn H. Hjertager, University of Stavanger, Stavanger, Norway*
*Dr. Tron Solberg, Aalborg University, Esbjerg, Denmark*

## Introduction

The current study deals with the prediction of different chemical and physical processes occurring within an entrained-flow coal gasifier using computational fluid dynamics (CFD). A gasifier is a chemical reactor in which a feedstock (fuel) is combusted in the presence of oxygen, usually at an elevated pressure and at high temperatures. The process in which combustible materials are partially oxidized (partially combusted) is known as gasification. Partial oxidation of combustible material generally takes place in an oxygen-lean environment at temperatures ranging from 760 to 1540°C. The synthesis gas (Syngas) produced as a result of the gasification process is composed primarily of a combustible mixture of carbon monoxide and hydrogen (water and carbon dioxide being the other major constituents).

The abundant availability of coal coupled with the flexibility of feedstock usage, fluctuations in natural gas prices and increased environmental concerns make gasification an attractive alternative to conventional thermal power plant solutions for the energy requirements of tomorrow. Power plants using the Integrated Gasification Combined Cycle (IGCC) technology in particular, are being developed worldwide to utilize conventional feedstock such as coal in a cleaner and more efficient manner. Their intended use in gasification technology based power plants are expected to bring about improvements in reliability, emission control, efficiency and feedstock flexibility. The feedstock used for a gasification system could be coal, petroleum coke, biomass, heavy oil or even natural gas.

## Key Methods

Gasification is a complex physical and chemical process involving multiphase flow, heat transfer, mass transfer and chemical reactions. When the feedstock is introduced into the gasifier at a high temperature, a series of physical and chemical processes occur within the reactor. The volatile content in coal particles is devolatilized and the resulting char is burned or gasified. Simultaneously, the gaseous products from the gasification of the coal particles will be ignited or, react with other species based on the surrounding environment and intrinsic kinetics mechanisms [3]. These processes cannot be modeled in their entirety due to their complexity and the heterogeneity of coal. Based on experiments and simplified mechanisms, this process can be further divided into simple sub-processes such as inert heating, devolatilization, combustion, etc.

The CFD model in the present work has been developed to predict the performance of an entrained-flow laboratory scale gasifier described in [9] and elsewhere ([1]). This type of gasifier is widely used for coal gasification. Due to the short gas residence time within the entrained-flow gasifier, the temperature needs to be high and the coal needs to be pulverized to ensure high carbon conversion of the feedstock. The coal feed is injected in co-current flow with the oxidant which rapidly heats up the coal and reacts with it.

Even though extensive work has been carried out in the CFD modeling of gasification processes, for instance in [2], [3], [8] and [10], there are barely any studies that have extensively compared with experimental data species evolution and mole fractions at different locations within the gasifier. For this study, one type of feedstock has been considered (Utah bituminous coal), and this model will later be applied for other types of feedstock. For the proximate analysis of coal, the four following pseudocomponents are utilized: ash, moisture, fixed carbon and volatile matter. Moisture is released in the initial process of vaporization, followed by the release of volatiles from the coal during the devolatilization process, which produces various gas species. The fixed carbon component is consumed in the combustion and gasification reactions.

### *Multiphase Flow*

In an entrained-flow gasifier, the gas flow carries the coal particles and the flow regime can be considered dilute. This is because the volume occupied by the particles is very small when compared to the gas flow volume, and hence the particle-particle interactions are negligible. The discrete phase method (DPM) in FLUENT has been applied to model the multiphase flow. The use of this method is justified when the flow is dilute, that is, the particle volume fraction is less than 10%. Another restriction of the method is that the ratio of the mass flow rate of the solid particles to that of the gas should be less than or equal to one. Using the DPM technique, the particle trajectories in the multiphase flow can be computed (using a Lagrangian description) as well as mass and energy transfer between the particle and gas phases. In other words, the continuous or gas phase is handled with an Eulerian formulation and the discrete or particle phase is handled with a Lagrangian one.

The gasification process can be broken down into the following simple sub-processes for ease of modeling. Coal particles are subject to the following sub-processes during gasification [8].

### *Injection and Passive Heating*

The coal particles are injected in the gasifier using a carrier gas. The particle sizes are assumed to fit a Rosin-Rammler distribution. The coal particles are heated till the vaporization (and later, devolatilization) temperature is reached. This sub-process is characterized by an absence of mass transfer or chemical reactions.

### *Release of Moisture, Particle Swelling and Devolatilization*

Moisture is released when the coal particles cross a threshold temperature known as the vaporization temperature. This moisture is then added to the gas continuity equation and the species transport equation as a source. The energy for the latent heat of vaporization is extracted from the gas phase. The particle diameter changes during devolatilization according to the swelling coefficient which is defined by the user in FLUENT [5]. A phenomenological model is built based on data from lab experiments that characterize the coal. This stoichiometry determining model predicts the yields of major gas components while preserving strict elemental balance. The main species included in this devolatilization model are CO, $H_2$, $CO_2$, $H_2O$, $O_2$, CH4, $H_2S$ and $N_2$. The devolatilization rate is determined from the two-competing rates Kobayashi model. The Kobayashi model is given by the following equation,

$$\frac{dY}{dt} = \frac{d(Y_1 + Y_2)}{dt} = k_{d1}(Y_1^* - Y_1) + k_{d2}(Y_2^* - Y_2) \tag{1}$$

Where, $k_{d1} = A_1 \exp(-E_1 / RT_p)$ and $k_{d2} = A_2 \exp(-E_2 / RT_p)$ are the two competing rates.

From [2] and [4], for oxygen lean conditions and Utah bituminous coal which was used in the CFD model, we obtain

$$volatiles + 0.479O_2 \xrightarrow{k1} 1.445CO + 1.757H_2 + 0.012H_2S + 0.032N_2 \quad (2)$$

Where, $C_{m11}H_{m12}O_{m13}N_{m14}S_{m15}$ represents the volatiles released by the coal during devolatilization.

### *Char Combustion and Gasification*

After the volatiles have been released, gasification and combustion reactions take place till all the char is consumed or the particles exit the gasifier. The chemical reactions during this stage include the oxidation of char and the reaction of char with carbon dioxide, water and hydrogen. Carbon monoxide and hydrogen are produced during the gasification of char particles. The performance of a gasifier is determined primarily by the char gasification reaction. The following reactions have been included in the model:

$$C + \frac{1}{2}O_2 \xrightarrow{k2} CO \quad (3)$$

$$C + H_2O \xrightarrow{k3} CO + H_2 \quad (4)$$

$$C + CO_2 \xrightarrow{k4} 2CO \quad (5)$$

$$C + 2H_2 \xrightarrow{k5} CH_4 \quad (6)$$

The reaction rate for these heterogeneous reactions is given by the Arrhenius relation $k_s = A_c \exp(-E_c / RT_p)$

### *Gas Phase Reactions*

The continuous phase is modeled by using global reactions to describe gas phase chemistry. The following reaction paths have been considered for this model:

$$CH_4 + \frac{1}{2}O_2 \xrightarrow{k6} CO + 2H_2 \quad (7)$$

$$H_2 + \frac{1}{2}O_2 \xrightarrow{k7} H_2O \quad (8)$$

$$CO + \frac{1}{2}O_2 \xrightarrow{k8} CO_2 \quad (9)$$

$$CH_4 + H_2O \underset{k9b}{\overset{k9f}{\rightleftarrows}} CO + 3H_2 \quad (10)$$

$$CO + H_2O \underset{k10b}{\overset{k10f}{\rightleftarrows}} CO_2 + H_2 \quad (11)$$

A finite-rate/eddy-dissipation model is used to account for the interaction between chemistry and flow turbulence. In this model, the effective reaction rate is defined by taking the minimum of the chemical reaction rate and the turbulent mixing rate, in other words

$$R_{eff} = \min\left[ A_i T_g^n \exp\left( -\frac{E_i}{RT_g} \right) [X]^a [Y]^b , C_\mu \rho \frac{k}{\varepsilon} \min(m_{fu}, m_{ox}/\phi) \right] \tag{12}$$

Where, $A_i$ is the pre-exponential coefficient, $T_g$ is the gas temperature, $E_i$ is the activation energy, $a$ and $b$ are global reaction exponents for the species $X$ and $Y$, $m_{fu}$ and $m_{ox}$ are the mass fractions of the fuel and oxidant and $\phi$ is the stoichiometric ratio. The values of the pre-exponential coefficient and the activation energies for the numbered chemical reactions to determine the reaction rates are given in the table below.

**Table 1**. Rate Coefficients for Heterogeneous and Gas Phase Reactions

| Reaction Rate | A | n | E (J/kmol) | Reference |
|---|---|---|---|---|
| k1 | $2.119 \times 10^{11}$ | 0 | $2.027 \times 10^{8}$ | [5] |
| k2 | 0.86 | - | $1.49 \times 10^{8}$ | [6] |
| k3 | 1.33 | - | $1.47 \times 10^{8}$ | [1] |
| k4 | 4.40 | - | $1.62 \times 10^{8}$ | [1] |
| k5 | $1.18 \times 10^{-5}$ | - | $1.49 \times 10^{8}$ | [6] |
| k6 | $3.0 \times 10^{8}$ | -1 | $1.26 \times 10^{8}$ | [10] |
| k7 | $6.8 \times 10^{15}$ | 0 | $1.68 \times 10^{8}$ | [10] |
| k8 | $2.2 \times 10^{12}$ | 0 | $1.67 \times 10^{8}$ | [10] |
| k9f | $4.4 \times 10^{11}$ | 0 | $1.68 \times 10^{8}$ | [10] |
| k9b | $5.12 \times 10^{-14}$ | 0 | $2.73 \times 10^{4}$ | [10] |
| k10f | $2.75 \times 10^{10}$ | 0 | $8.38 \times 10^{7}$ | [10] |
| k10b | $2.65 \times 10^{-2}$ | 0 | $3.96 \times 10^{3}$ | [10] |

***Flow and Heat Transfer Models***

The standard k-ε model was used for modeling turbulence in the continuous phase, and stochastic tracking was implemented for the steady-state three dimensional model of the gasifier to simulate particle motion and turbulent dispersion. The SIMPLE algorithm was used to implement the pressure-velocity coupling with second-order upwind discretization. The coal particles were assumed to fit a Rosin-Rammler size distribution.

Conductive, convective and radiative heat transfer was assumed between the gases, particles and reactor walls. The scattering property of the particles due to radiation is assumed to be isotropic and the standard P1 model has been used for the purpose of modeling radiative heat transfer.

## Model Details

The gasifier geometry is as detailed in the lab scale gasification experiments of [9]. The length of the gasifier is 1.37m and the nominal diameter is 0.2m. Gambit was used to generate the mesh for the 3D geometry of the gasifier which consisted of 68070 tetrahedral elements. The discrete phase method in FLUENT 12.1 was used for the CFD simulation.

As given in [9], the design coal feed rate is 24.5kg/hr. The mass ratio of coal to oxygen to steam is 1.0:0.91:0.27. There are two streams entering the gasifier, the primary stream consists of a pre-mixture of oxygen and argon with coal, and the secondary stream consists of a mixture of steam and helium. The primary stream was preheated to 366K and the secondary stream to 430K. The argon mass flow rate is 3.31kg/hr and that of helium is 0.076kg/hr.

The coal particles were assumed to have a Rosin-Rammler size distribution with a minimum size of 4 micro meters and a maximum of 76, with an average of 42 micro meters. The spread parameter was 1.2 and 10 discrete particle sizes were considered.

## Results and Discussion

The gasifier simulations were run using FLUENT 12.1 on a 64-bit LINUX cluster utilizing 8 processors (2.3 GHz). An average run took close to 48 hours for over 40,000 iterations to obtain solution convergence. The temperature profile and mole fraction distributions for carbon dioxide, carbon monoxide and hydrogen within the gasifier obtained from the FLUENT runs were compared with the experimental results from [9]. The entrained-flow gasifier was preheated to a temperature of between 1400 and 1700K with a methane pre-burner before the coal particles were injected, hence this temperature was patched to the gas phase before the simulation run was started.

The flow field velocity distribution within the gasifier shows strong downward flow at the inlet and outlet and regions of recirculation near the top and side walls of the gasifier which is caused due to the high velocity of injection of the coal and oxygen. Below the recirculation regions, there is flow akin to plug flow – these classical flow characteristics have been observed in other gasification studies. The axial velocity profile is illustrated in figure 1. below.

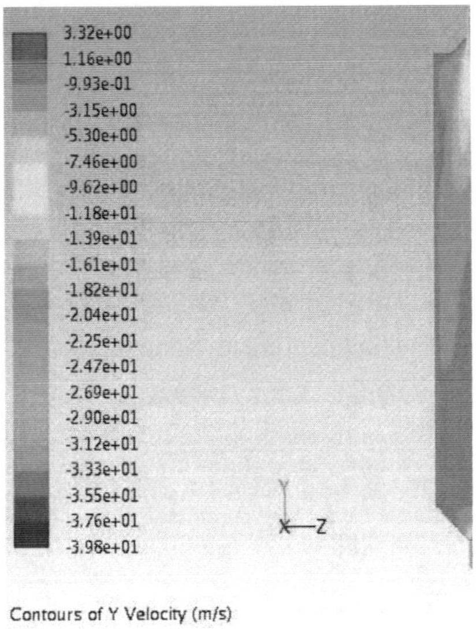


**Figure 1**. Axial velocity distribution

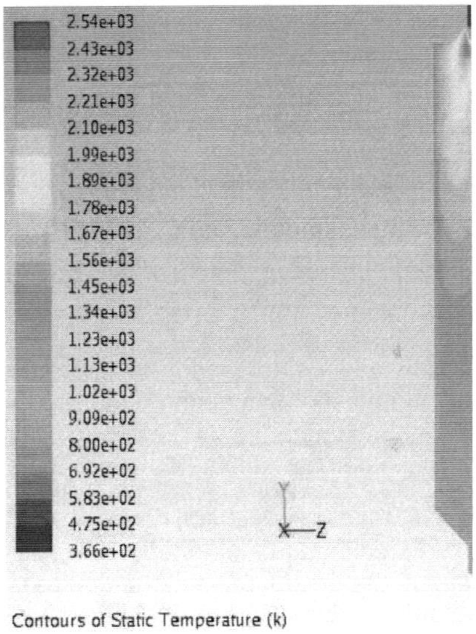


**Figure 2**. Temperature distribution

Figure 2. illustrates the temperature distribution within the gasifier obtained from the simulation. It is seen that the highest temperature (about 2540K) is observed at the center, near the top of the gasifier (less than 0.1m from the top walls).which is mainly due to the process of combustion of the coal and oxidant. The combustion zone extends about 0.1m in the axial direction and below this zone the gas temperature decreases gradually as we move towards the outlet. This decrease in temperature is due to the gasification process which is an endothermic reaction consuming heat. The simulated peak temperature contour resembles the shape of a flame.

If we look at the mole fractions for the different species, we see that there is good agreement with both the values and radial trends for carbon dioxide with the experimental results. We see that the maximum deviation from the experimental values in this case occurs close to the inlet, and close to the axis of the gasifier. This variation could be attributed to high turbulence intensity in those regions, the physics of which the k-ε turbulence model captures weakly (especially in the presence of strong swirl). The difference in carbon dioxide mole fractions between the experimental data and the CFD predictions is a little over 6% at the outlet. This can be seen in figure 3.

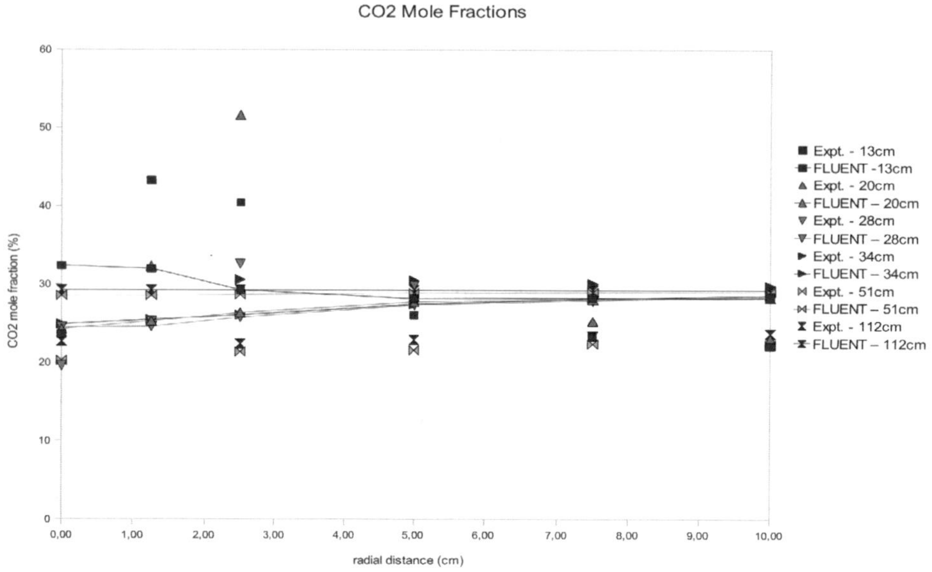


**Figure 3**. Radial profile of carbon dioxide mole fraction

In the case of carbon monoxide we see a large under-prediction of mole fraction values from the CFD simulation, even though by and large it captures the trends exhibited by the experimental data. The difference in carbon monoxide mole fractions between the experimental data and CFD predictions ranges between 15 and 18%. This is illustrated in figure 4.

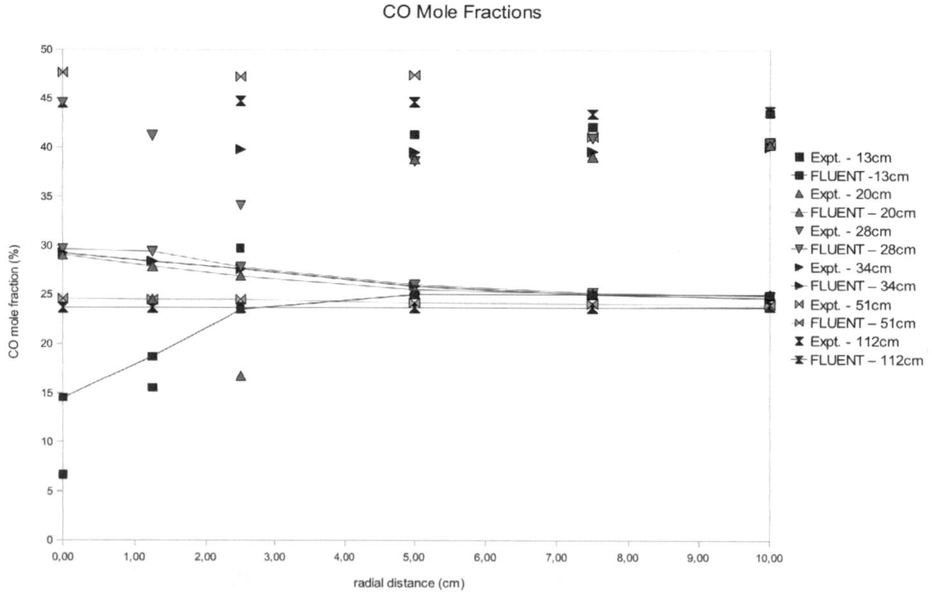


**Figure 4**. Radial profile of carbon monoxide mole fraction

In the case of hydrogen, it is seen that the experimental trends are captured, but the model over-predicts the mole fractions for different radial positions within the gasifier. The difference between the experimental and predicted mole fractions at the outlet of the gasifier varies between 8 and 15%. The various profiles are depicted in figure 5.

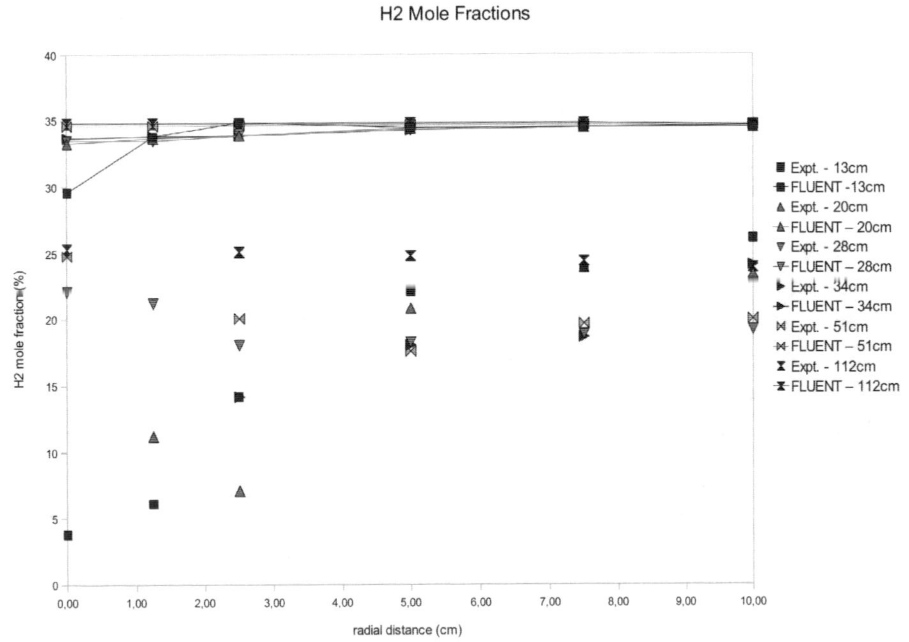


**Figure 5**. Radial profiles of hydrogen mole fractions

## Conclusion

When compared with the experimental data from [9], it is seen that the CFD model captures the radial trends for the molar fractions of carbon dioxide, carbon monoxide and hydrogen quite well. The absolute values of mole fractions at different radial locations within the gasifier are captured well for carbon dioxide, but the current model under-predicts the carbon monoxide mole fraction and over-predicts the hydrogen mole fraction.

Improvements are expected in the results from model refinements relating to the chemical kinetics, detailed modeling of the gasifier inlet, numerical schemes including the turbulence model, inclusion of swirl, radiation models used and mesh refinement. The best agreement with experimental data has been obtained for the mole fraction of carbon dioxide and the worst with carbon monoxide. This issue has to be looked into further and work will be done to bridge the gap between simulation results and experimental data.

## Acknowledgements

I would like to thank Statoil Research Center Trondheim (Norway) for providing financial support for this study.

## References

[1] Brown, B.W., Smoot, L.D., Smith, P.J., and Hedmen, P.O. (March 1988), "Measurement and Prediction of Entrained-Flow Gasification Processes," *AIChE Journal*, **34(3)**, pp. 435-446.

[2] Chen, C., Horio, M., and Kojima, T. (2000), "Numerical simulation of entrained flow coal gasifiers. Part I: modeling of coal gasification in an entrained flow gasifier," *Chemical Engineering Science*, **55**, pp. 3861-3874.

[3] Choi, Y.C., Li, X.Y., Park, T.J., Kim, J.H., and Lee, J.G. (2001), "Numerical study on the coal gasification characteristics in an entrained flow coal gasifier," *Fuel*, **80**, pp. 2193-2201.

[4] EPRI AP-4509, Project 2221-1, Final Report (April 1986), "Use of Lignite in Texaco Gasification-Based-Combined-Cycle Power Plants."

[5] FLUENT 6.3 Theory Guide.

[6] Govind, R., and Shah, J. (1984), "Modeling and Simulation of an Entrained Flow Coal Gasifier," *AIChE Journal*, **30 (1)**, pp. 79-92.

[7] Jones, W.P., and Lindstedt, R.P. (1988), "Global Reaction Schemes for Hydrocarbon Combustion," *Combustion and Flame*, **73**, pp. 233-249.

[8] Shi, S.-P., Zitney, S.E., Shahnam, M., Syamlal, M., and Rogers, W.A. (2006), "Modelling coal gasification with CFD and discrete phase method," *Journal of the Energy Institute*, **79(4)**, pp. 217-221.

[9] Soelberg, N.R., Smoot, L.D., and Hedman, P.O. (1984), "Entrained flow gasification of coal – 1. evaluation of mixing and reaction processes from local measurements," *Fuel*, **64**, pp. 776-781.

[10] Watanabe, H., and Otaka, M. (2006), "Numerical simulation of coal gasification in entrained flow coal gasifier," *Fuel*, **85**, pp. 1935-1943.

# (624d) On the Modeling of a Single-Stage, Entrained-Flow Gasifier Using Aspen Custom Modeler

Job Kasule[1], Richard Turton[2], Debangsu Bhattacharyya[1] and Stephen E. Zitney[3], (1)Department of Chemical Engg., West Virginia University, Morgantown, WV, (2)Chemical Engineering, West Virginia University, Morgantown, WV, (3)Collaboratory for Process and Dynamic Systems Research, National Energy Technology Laboratory, U.S. Department of Energy, Morgantown, WV

Coal-fired gasifiers are the centerpiece of integrated gasification combined cycle (IGCC) power plants. The gasifier produces synthesis gas that is subsequently converted into electricity through combustion in a gas turbine. Several mathematical models have been developed to study the physical and chemical processes taking place inside the gasifier. Such models range from simple one-dimensional (1D) steady-state models to sophisticated dynamic 3D computational fluid dynamics (CFD) models that incorporate turbulence effects in the reactor. The practical operation of the gasifier is dynamic in nature but most 1D and some higher-dimensional models are often steady state. On the other hand, many higher order CFD-based models are dynamic in nature, but are too computationally expensive to be used directly in operability and controllability dynamic studies. They are also difficult to incorporate in the framework of process simulation software such as Aspen Plus Dynamics. Thus lower-dimensional dynamic models are still useful in these types of studies.

In the current study, a 1D dynamic model for a single-stage, downward-firing, entrained-flow GE-type gasifier is developed using Aspen Custom Modeler® (ACM), which is a commercial equation-based simulator for creating, editing, and re-using models of process units. The gasifier model is based on mass, momentum, and energy balances for the solid and gas phases. The physical and chemical reactions considered in the model are drying, devolatilization/pyrolysis, gasification, combustion, and the homogeneous gas phase reactions. The dynamic gasifier model is being developed for use in a plant-wide dynamic model of an IGCC power plant.

For dynamic simulation, the resulting highly nonlinear system of partial differential algebraic equations (PDAE) is solved in ACM using the well-known Method of Lines (MoL) approach. The MoL discretizes the space domain and leaves the time domain continuous, thereby converting the PDAE to a differential algebraic equation (DAE) system with respect to time. The DAE system is solved using a variable-step implicit Euler integrator. For steady-state simulations, the set of nonlinear algebraic equations are solved using a Newton-type method.

In this presentation, preliminary results from the steady-state non-isothermal gasifier model will be reported. Comparisons of the results from the gasifier model to available pilot plant data, industrial gasifier data, and other published models will be made. Sensitivity studies will be presented for different types of coal and operating conditions.

See more of this Session: Advances in Gasification Research II
See more of this Group/Topical: Topical E: High Temperature Environmentally Sustainable Energy Processes (sessions joint with the Environmental Division)

# SYSTEMATIC PERFORMANCE ANALYSIS OF HYDROGEN AND POWER CO-PRODUCTION SYSTEM BASED ON $CO_2$ SORPTION ENHANCED COAL GASIFICATION

*Xun Wang* *Wendong Tian* *Yunhan Xiao*

Institute of Thermophysics Engineering, Chinese Academy of Sciences, Beijing, China

## Abstract

This paper proposed a kind of hydrogen production / power generation system and investigates its performance from the point of view of chemical energy conversion and green house gas (GHG) emissions. This system had the advantage of nearly zero emission. It was based on $CO_2$ sorption enhanced coal gasification technique and SOFC power generation. Syngas of gasification was mainly composed of 60-90% hydrogen and 10-40% methane. Contents of $CO_2$ and CO can be lower than 1%. In order to investigate the performance of gasifier, which in fact, is far beyond chemical equilibrium, a gasification kinetics unit model was developed based on char-gas, gas-gas, absorbent-$CO_2$ reaction kinetics. This model can predict the gas composition, syngas yield, solid(unconverted char, ash, unconverted absorbent, limestone) content in bed materials. Gasification operation conditions (temperature, pressure,) will determine carbon conversion ratio, gases qualities and quantity of gasification process. Syngas yield and gas content will then determine performance (thermal efficiency, $CO_2$ emissions) of co-production system. Systematic study was done by way of flowsheet simulation and parameters sensitive studies. Influence of key operation conditions on units and system can then be analyzed. Results provided ways to system optimization.

## Introduction

$CO_2$ Sorption Enhanced Coal Gasification [1-3] integrates $H_2$ production and $CO_2$ separation in one reactor. Due to continuous sorption of $CO_2$ by sorbent, $H_2$ yield is prospective to be increased greatly and pure $CO_2$ can be captured at relatively low cost.

The gasification is divided into two reactors: a gasifier with $CO_2$ sorption and a sorbent regenerator. They are connected by recycling absorbent/heat carrier. In the first reactor, carbon is converted to syngas and meanwhile, $CO_2$ released from this process is absorbed by absorbent. Syngas is mainly composed of 60-90% hydrogen and 10-40% methane. Contents of $CO_2$ and CO can be lower than 1%. In second reactor, used absorbent is calcined. The regenerated absorbent is recycled to the gasifier to absorb $CO_2$ and to provide heat of endothermic reactions. Thus with calcination process, $CO_2$ can be released and separated at nearly pure state.

The influence of temperature and pressure on gasifier performance is key issue concerned in many references. It has been studied either by theoretical analyze [1-5] or by experiment [4,5,6]. Up to now, most theoretical study are based on chemical equilibrium [1-5,7]. Equilibrium gas compositions from C/coal–$H_2O$–CaO reaction systems were calculated based on the assumption of gibbs energy minimization. Results showed that CO and $CO_2$ levels were lower, and $H_2$ was higher than those in the general hydrocarbon gasification. CO and $CO_2$ decreased with increasing pressure, and reached to approximately zero at higher pressure. $CO_2$ fixation by CaO is a major contributor to the CO reduction and the $H_2$ increase. $H_2$ was the primary gas, with a small amount of $CH_4$. These work are very important for the study of the influence of gasification operation parameters—temperature, pressure, ration of calcium oxide to carbon(CaO/C), ratio of steam to carbon(S/C).

In fact, due to the lower operation temperature(600℃-1000℃) , gasification process is far beyond chemical equilibrium. Coal conversion ratio and product distribution are heavily dependent on chemical kinetics and hydrokinetics. Temperature and pressure are two main factors that have strong effect on chemical kinetics. So it is necessary to study the gasification process by means of chemical kinetics. Reference [5] set up a model to simulate the dynamic performance of gasifier from the point of view of chemical kinetics.

On the basis of this model and coal gasification simulation knowledge in literature[8,9], a gasification kinetics steady state model was developed based on kinetics mechanism of char-gas, gas-gas, and carbonation reaction. It was used to predict the parameters of gas phase(gas composition, syngas yield), solid phase(unconverted char, ash, unconverted absorbent, limestone content in bed and discharge). Carbon conversion ratio, gases qualities and quantity of gasification process were determined by gasification operation conditions (temperature, pressure, calcium oxide to carbon ratio, steam to carbon ratio).

A $H_2$ and power co-production system was designed. This system had the advantage of nearly zero emission. It was based on $CO_2$ sorption enhanced gasification and fuel cell power generation cycle. There was complicated mass and heat transfer between different parts. Systematic study was done by flowsheet simulation and parameters sensitive studies. Influence of key operation conditions on units and system were analyzed.

## Reactor Steady state modelling for $CO_2$ Sorption Enhanced Coal Gasification

It was assumed that gas phase was at the state of plug flow. Solid phase was totally mixed with homogenous component distribution. Solid-gas reactions and gas –gas reaction considered in this paper were listed below.

- $C+H_2O \leftrightarrow CO+H_2$
- $C+CO_2 \leftrightarrow 2CO_2$
- $C+2H_2 \leftrightarrow CH_4$
- $CaO+CO_2 \leftrightarrow CaCO_3$
- $CO+H_2O \leftrightarrow CO_2+H_2$
- $CH_4+H_2O \leftrightarrow CO+3H_2$

### *Mass and energy balance of reactor*

It was assumed that devolatilization was completed instantaneously. Thus the gasification process was assumed to be composed of a pyrolysis sub-reactor and gasification sub-reactor. They can be thought connected in serial (fig.1). The prediction of the final product yields of coal decomposition was referred to reference [10]. Results were listed in table 1.

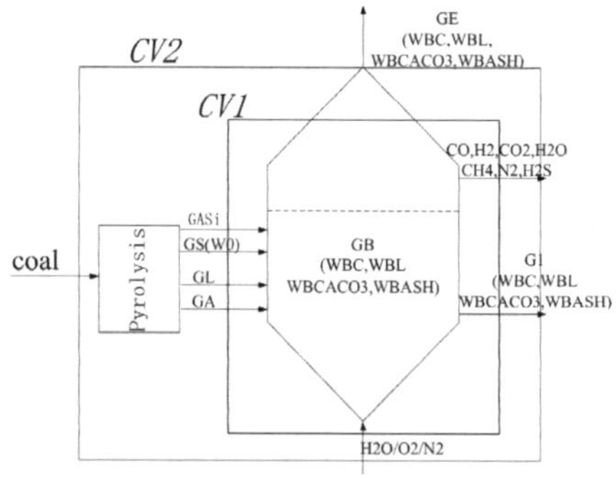

Fig.1 Control volume of gasification

The product of pyrolysis was used as the boundary condition of gasification sub-reactor. Mass balance of Char, $CaO$,$CaCO_3$, ash were established in control volumn-CV1. Along the height of the reactor, cells were divided in which mass balances model of each gas components-- $CO$, $CO_2$, $H_2$, $H_2O$, $CH_4$, $H_2S$, $N_2$ was set up.

With the assumption that temperature was homogeneous in whole reactor, pyrolysis sub-reactor and gasification sub-reactor were put together to model energy balance(CV2). The operation temperature or heat released was calculated at given feed coal mass flow, steam flow, sorbent flow and reactor operation pressure.

In the prediction model, the inputs included coal properties, steam flow, flow of CaO and $CaCO_3$, bed weight, reactor total pressure, inlet superficial gas velocity. The outputs were conversion ratio of carbon and CaO, Solids flow and composition, syngas flow and composition, operation temperature or heat released.

***Simulation model***

1) element balance

C and Ca balance of gasification sub-reactor at T=850℃ P=20bar were listed in table 1. The similar balance can be set up to element H,O and S.

Table 1 Carbon Balance(shen-mu coal[5], T=850℃ P=20bar)

| Feedstock | | Pyrolysis product（wt,%） | | | |
|---|---|---|---|---|---|
| Coal flow，g/s | 5 | C | 0.58820 | $CH_4$ | 0.03680 |
| Ration(CaO/coal)，g/g | 3.04 | $H_2O$ | 0.11370 | $N_2$ | 0.00884 |
| CaO flow，g/s | 15.20 | $H_2$ | 0.02900 | $H_2S$ | 0.00232 |
| Ratio(Steam/coal)，g/g | 1.96 | $CO_2$ | 0.04101 | ASH | 0.12570 |
| Steam flow，g/s | 9.8 | CO | 0.048014 | | |
| Inputs | | | | | |
| C, mol/s | 0.26970 | Ca, mol/s | 0.27143 | | |
| Outputs | | | | | |
| Gas phase | | Solid phase | | | |
| Composition（mol，%） | | Composition（wt，%） | | | |
| $CO_2$ | 0.729 | C | 0.47 | | |
| CO | 1.310 | CaO | 30.4 | | |
| $CH_4$ | 7.230 | $CaCO_3$ | 62.2 | | |
| Syngas flow, mol/s | 0.36083 | Discharge flow，g/s | 23.3 | | |
| Carbon in gas phase, mol/s | 0.03340 | Carbon in solid phase, mol/s | 0.23618 | | |
| Ca in gas phase, mol/s | 0. | Ca in solid phase, mol/s | 0.27129 | | |
| Total molar flow of carbon, mol/s | 0.26963 | Total molar flow of Ca, mol/s | 0.27129 | | |
| Error，% | 0.02600 | Error，% | 0.05200 | | |

2）test with experiments or running results

Simulation results were compared with experiment and plants results. Results were shown in tables 2-3.

Table 2: Comparison Between Fluidized Gasification[11]

| | Winkler-UR lignite | | | UGAS-washed | | |
|---|---|---|---|---|---|---|
| Molar fraction (dry basis，%) | E | C | Error.% | E | C | Error,% |
| $H_2$ | 35.3 | 38.93 | 10 | 42.2 | 43.24 | 2.5 |
| $CO_2$ | 13.3 | 11.47 | 14 | 28.6 | 23.39 | 18.2 |
| CO | 48.2 | 42.80 | 11 | 25.6 | 29.63 | 15.8 |
| $CH_4$ | 1.8 | 5.63 | 213 | 2.9 | 2.28 | 21.1 |
| $N_2$ | 0.9 | 0.90 | 0.00 | 0.6 | 0.55 | 8.3 |
| Xc | 57-88 | 70.80 | | 89.7 | 98. | 9.9 |

E: experiment or plants results   C: simulation results   Xc: carbon conversion ratio

Table3  Comparison Between $CO_2$ Sorption Enhanced Gasification Experiment results[12]

| | T=650℃ | | T=700℃ | | T=800℃ | |
|---|---|---|---|---|---|---|
| molar fraction(dry basis，%) | C | E | C | E | C | E |
| $H_2$ | 74.13 | 77.20 | 84.6543 | 82.60 | 92.50 | 74.9 |
| $CO_2$ | 2.84 | 1.10 | 0.099 | 1.20 | 0.040 | 6.8 |
| CO | 0.065 | 0.80 | 0.0192 | 1.00 | 0.014 | 1.6 |
| $CH_4$ | 17.81 | 19.00 | 11.99 | 15.10 | 6.13 | 14.2 |
| $H_2$ flow，L/gcoal | 0.21 | 0.3 | 0.38 | 0.33 | 1.03 | >0.4 |

With the comparison, it can be found that simulation can show the basic characteristics of $CO_2$ sorption enhance gasification. But the prediction error were still high, especially to low content component, such as $CH_4$, CO,$CO_2$. The simulation needs to be improved by the following means.

- Correct semi-experience parameters of devolatilization model by experiment. This is helpful to decrease prediction error, especially to $CH_4$ composition.
- Adopt more comprehensive hydrokinetics model[8].
- Correct semi-experiment parameters of chemical kinetics model.
- Study reaction characteristics of $CO_2$ adsorption enhance gasification in its unique operation condition.

One possible reason for the high prediction error of $CO_2$ and CO content maybe the reaction rate of carbonation reaction adopted in this paper. It was higher than that of experiment and can be speedup evidently by temperature. Thus very low $CO_2$ and CO contents were obtained at higher temperature in simulation.

***Influence of temperature and pressure on gasification***

1) parameter used in this paper

- Xcoal-eg[1]: The conversion efficiency of coal to $H_2$ in pyrolysis and gasification.
- Xc: carbon conversion ratio
- $X_{CaO}$: CaO conversion ratio
- Effective gas production per unit cal: refer to flow sum of $H_2$ and $CH_4$ / coal flow

2) conversion ratio and syngas production

Fig 2.1-2.3 were equilibrium results at different temperature and pressure. In this simulation, gas compositions were assumed at the state of chemical equilibrium. The results were same with reference[5].But it was found that conversion value of carbon(did not shown in figures) and CaO were disagree with experiment. First it was impossible to achieve 100% carbon conversion ratio. Also CaO conversion ratio decreased with temperature. Coal to $H_2$ efficiency Xcoal-eg was not sensitive to temperature. This was also different with experiment results[12].

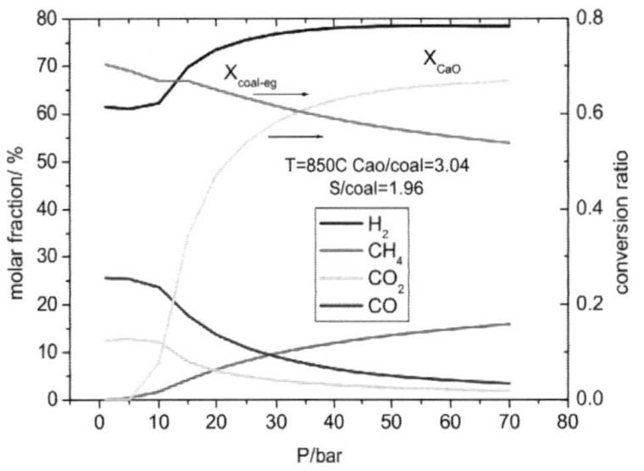

Fig.2.1 Equilibrium compositions at fixed temperature

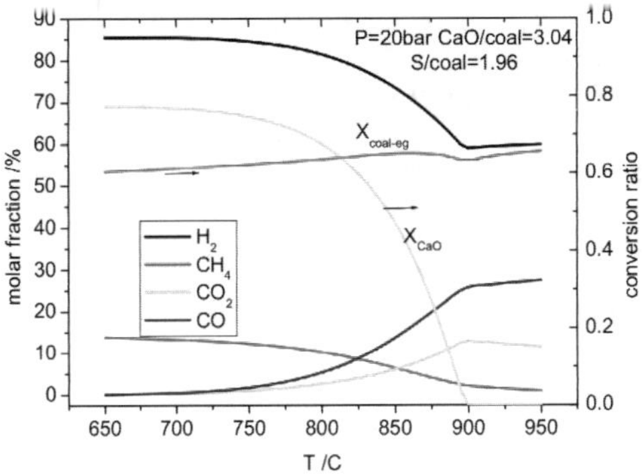

Fig.2.2 Equilibrium compositions at fixed pressure 20bar

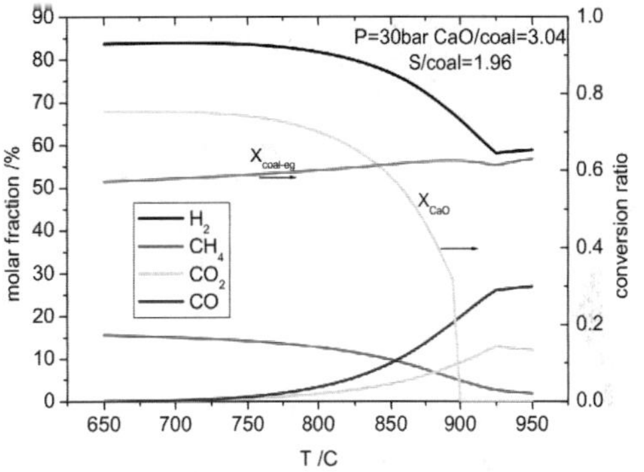

Fig.2.3 Equilibrium compositions at fixed pressure 30bar

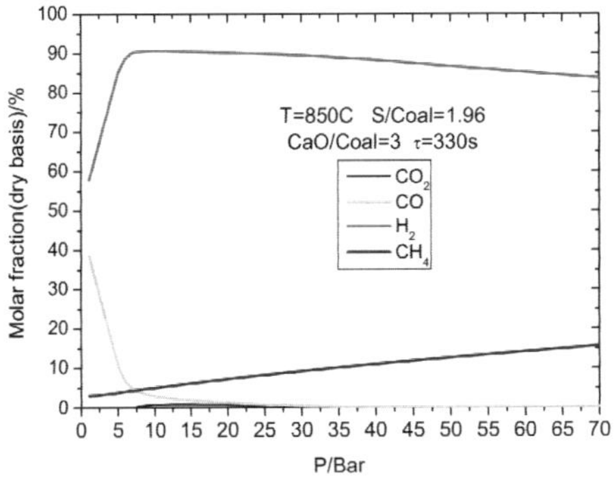

Fig2.4 Chemical kinetics gas composition at fixed temperature

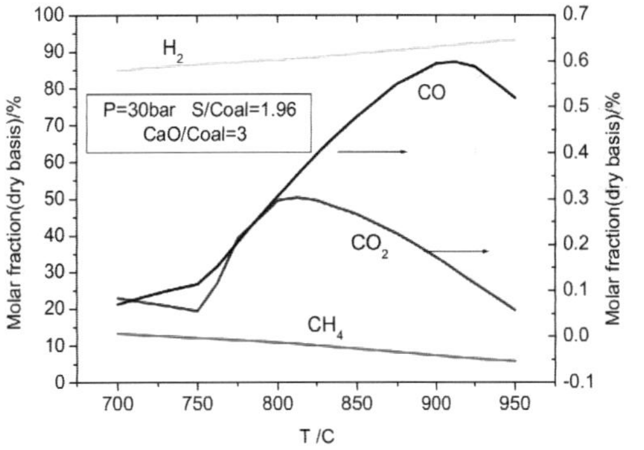

Fig2.5 Chemical kinetics gas composition at fixed pressure

Figs 2.4-2.8 were results by means of kinetics simulation. Results illustrated that temperature was an important factor for conversion ratio and effective gas production variation. With temperature increase, $CH_4$ content was decreased. But at given temperature, there is a pressure at which point coal to $H_2$ efficiency was maximal. With the increases of pressure, $H_2$ content was first increased and then decreased slowly. $CH_4$ would increase relative faster. Gas composition and conversion were not dependent on pressure as greatly as on temperature. These were coincident with experiment and reference[1,4,5,7,12]. In reference[5], $H_2$ composition was also increased with temperature. But in reference[12], though absolute flow of $H_2$ increased with temperature, at higher temperature, its content would decrease. The conflict may lies in the carbonation reaction kinetics used in simulation. Due to higher sorption ability, especially at higher temperature, the $CO_2$ and CO content calculated were much lower than that of experiment, which would lead to higher $H_2$ content.

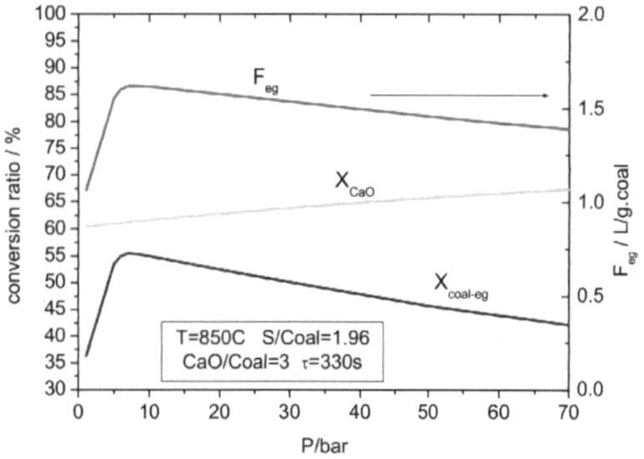


Fig2.6 Conversion ratio and effective gas yield($H_2$+$CH_4$) at fixed temperature

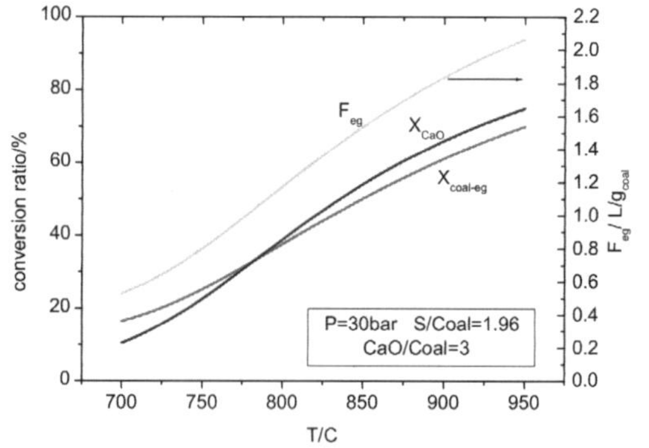


Fig2.7 Conversion ratio and effective gas yield($H_2$+$CH_4$) at fixed pressure

From gasification chemical kinetics simulation, it was concluded that increase temperature was advantage to increase $H_2$ production and decrease $CH_4$ content. But at higher temperature, the heat released in gasification was decreased(fig.2.8). At last, the reaction became endothermal with increased temperature. So temperature could not be so high in order to ensure reaction exothermal. Others factors that restrict temperature increase were eutectic problem of sorbent, decomposition of calcium carbonate, etc.

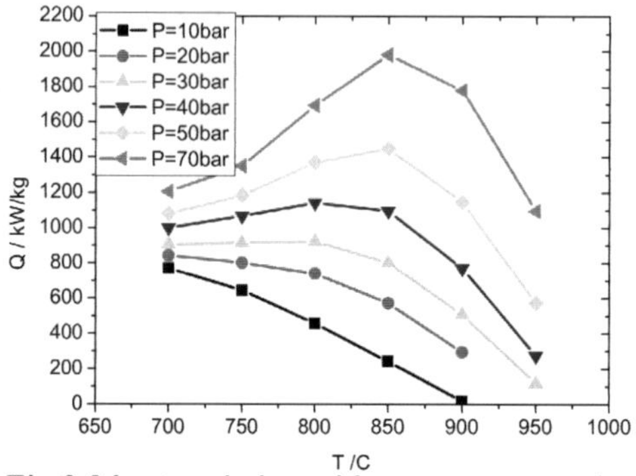


Fig.2.8 heat variation with temperature and pressure

3) one dimensional gas distribution along reactor height

Comparing figs.3.1-3.4, it showed that the higher the temperature was, the more evident the influence of pressure was. This was as the same tendency as reference [5]. The higher the pressure was, the more evident the flow change of $CH_4$ was. At given height and pressure, higher temperature corresponded to higher $CO_2$ and CO relative flow. At given height and temperature, higher pressure corresponded to lower $CO_2$ and CO relative flow. That means in order to obtain a

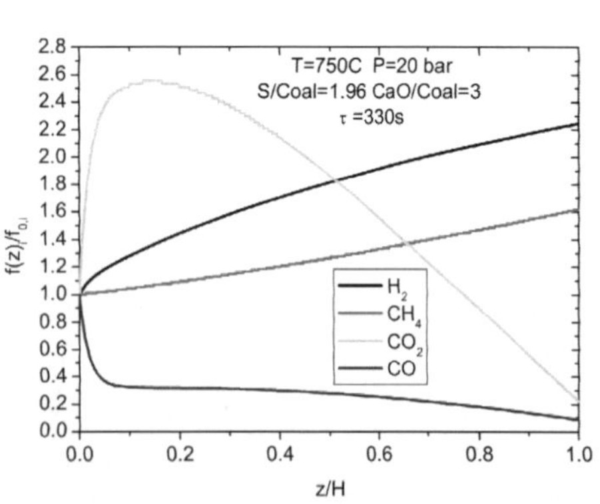


Fig.3.1 Relative flow variation of gas component along reactor height T=750℃ P=20bar

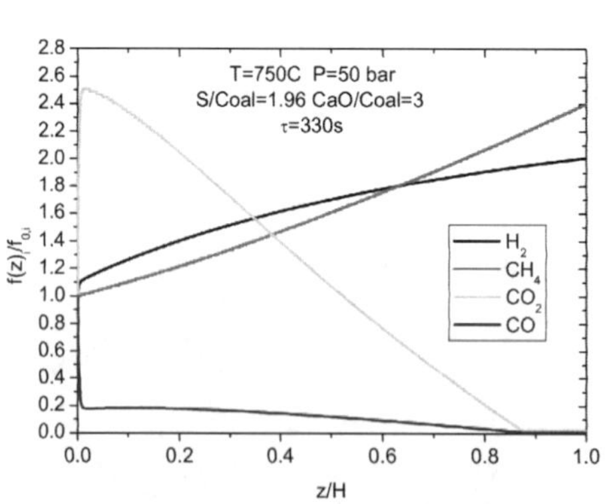


Fig.3.2 Relative flow variation of gas component along reactor height T=750℃ P=50bar

given $CO_2$ relative flow, at higher temperature and lower pressure, the reactor height should be higher. Here relative flow was defined as the flow of component i at height z divided by gasifier inlet flow of this component.

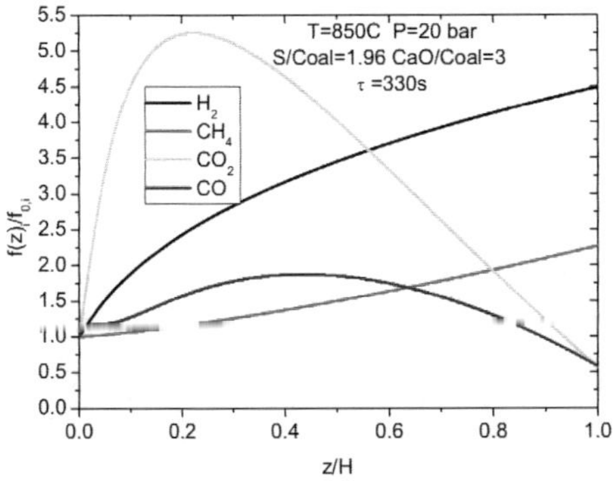


Fig.3.3 Relative flow variation of gas component along reactor height T=850℃ P=20bar

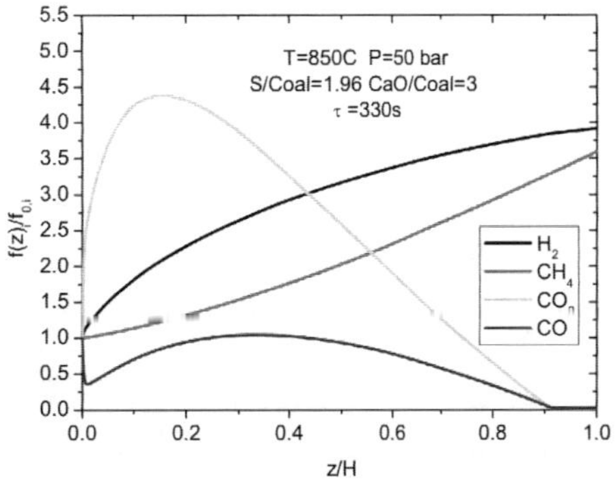


Fig.3.4 Relative flow variation of gas component along reactor height T=850℃ P=50bar

4）solid composition

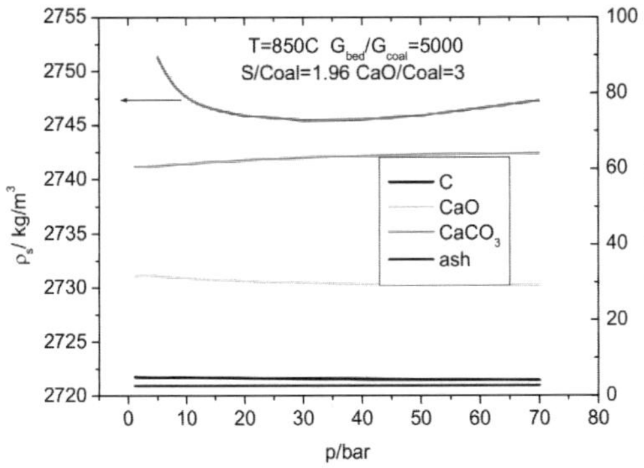


Fig.4.1 Solid composition and mixed density at fixed temperature

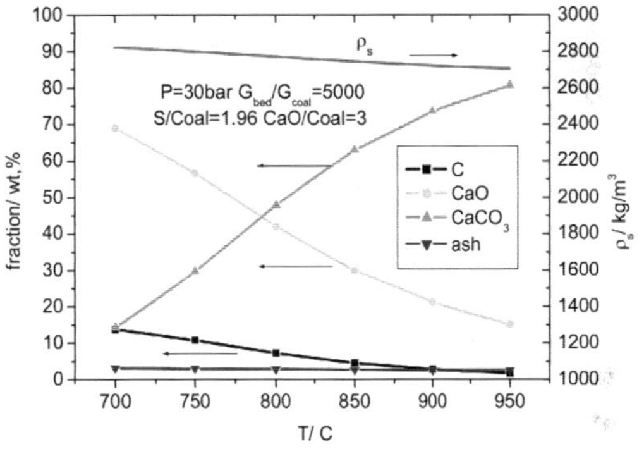


Fig.4.2 Solid composition and mixed density at fixed pressure

Fig.4.1 showed that the change of solid composition and density was not very sensitive to pressure. At pressure 30bar, the bed density was lowest. This means that if bed weight remained constant, the static height of bed at this point was highest. C, CaO and $CaCO_3$ content were influenced greatly by temperature(fig.4.2). But ash content changed very little with temperature. Density was decreased steadily with the increase of temperature.

5) reaction rate variation along reactor height

Along the height of reactor, reaction rates of carbonation reaction changed relatively slowly.

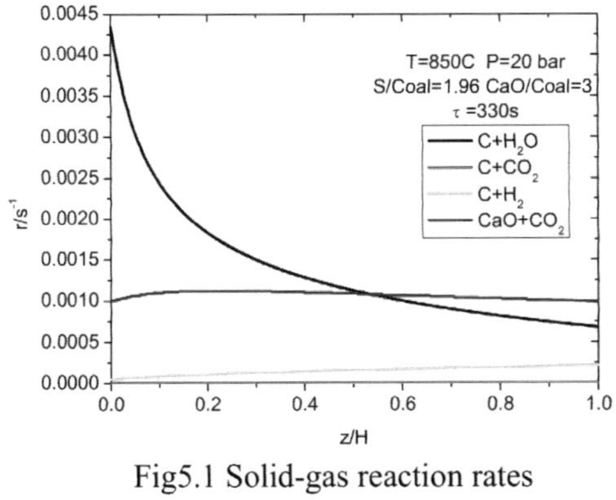


Fig5.1 Solid-gas reaction rates

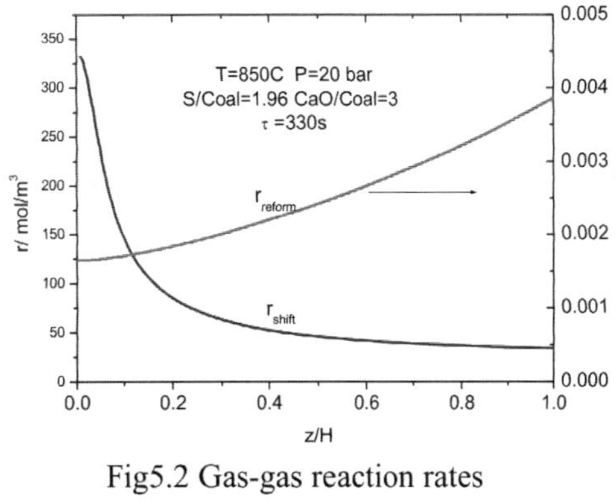


Fig5.2 Gas-gas reaction rates

At the inlet of reactor, the rates of steam gasification($C+H_2O$) reaction and shift reaction were high. But they declined quickly and then changed slowly at later part of the reactor. Rate of hydrogasification($C+H_2$) increased nearly linearly with height, which lead to increase of $CH_4$ content. While, methane steam reforming reaction also increased with height, which lead to steadily decomposition of methane.

## $H_2$ And Electricity Co-production

### *System configuration*

The description of $CO_2$ sorption enhanced gasification may be referred to reference [1-5]. Due to very low $H_2S$ content and high temperature, syngas was able to be fed to fuel cell directly without complicated sulfur removal and preheat process. Syngas after high temperature cleanup was fed to anode of a directly inner reforming SOFC. Compressed air heated up to 700℃ was fed to cathode of FC. Power was generated by electrochemistry reaction, and simultaneously, $CH_4$ could be converted to CO and $H_2$ by steam reforming reaction inside FC stack. With inner reforming, it was able to compensate $H_2$ consumed by electrochemistry reaction. On the other hand, $H_2$ and CO was reacted continuously, which was advantage to move chemical equilibrium of reforming reaction to the direction of $CH_4$ conversion. Some analysis showed that comparing with conventional methane steam reforming , reforming with electrochemistry reaction needed much smaller amount of steam, on condition that operation temperature, pressure and methane conversion ratio remained same. Stream containing $CO_2$, $H_2O$ and unconverted $CO,H_2,CH_4$ at anode outlet first released heat to air in heat exchanger and then was fed to $H_2$ separation unit. In this unit, the majority of $H_2$ was separated and exported as one product. Purge gas containing $H_2,CO,CO_2,CH_4,H_2O$ was recycled to regenerator. Cathode outlet stream also released heat in heat exchanger and then was expanded in a turbine to recovery power. And then it was fed to HRSG to generate superheated steam. Because the final gas product in regenerator was only $CO_2$ and $H_2O$, it was easy to capture $CO_2$ at nearly pure state. This made the system nearly zero emission. Super heated steam came from four sources. Being Generated by heat of gasifier and regenerator. Being generated in HRSG of cathode outlet. Being generated during the cooling process of regenerator outlet stream. Apart from provide super heated steam for gasification, the remaining steam flowed into a steam turbine to generate electricity which was output as another product.

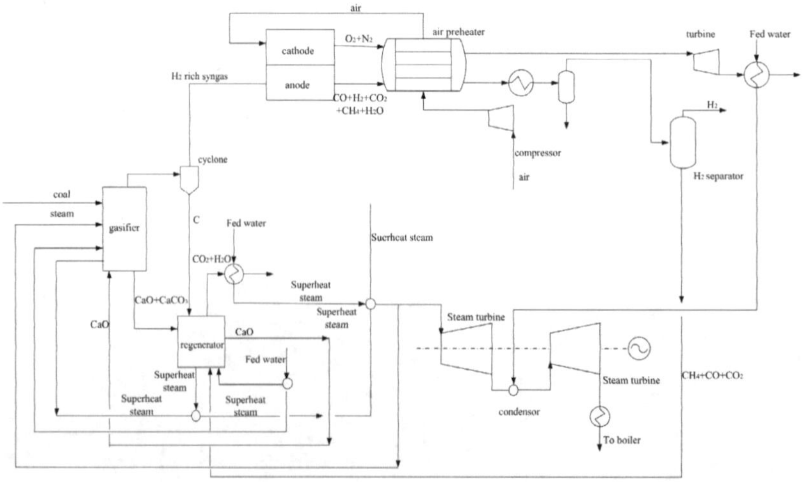


Fig.6 Simplified configuration of $H_2$, electricity co-production system

***Parameter definition***

- Total energy efficiency η: (thermal power of $H_2$ produced +electricity)/(LHV of coal*coal fed rate)
- Fuel utilization ratio U: $H_2$ converted in electrochemistry reaction/(flow sum of $H_2$,CO,and four times of $CH_4$ of inlet stream)
- Oxygen utilization ratio: oxygen consumed / inlet oxygen flow
- Relative conversion ratio of component i: flow of component i converted / inlet flow of this component, i= $CH_4$,CO
- Equivalent $CO_2$ emission per unit product: (Equivalent $CO_2$ emission in coal life cycle) /( thermal power of $H_2$ produced +electricity)

Coal life cycle includes coal mine exploitation, cleaning, transportation, gasification and syngas utilization. The calculation of equivalent $CO_2$ emission during coal mine exploitation, cleaning and transportation was interpreted in reference[13]. Emissions of gasification and syngas utilization were simulated in this paper.

- Equivalent $CO_2$ emission per unit $H_2$: (sum of $CO_2$ ,CO and $CH_4$ emission in coal gasification section) /( $H_2$ yield)

***Systematic simulation and results***

1) effect of stack reaction area

Change reaction area was one way to adjust fuel utilization rate and power output. Besides it had influence on stack temperature. In this paper, fuel cell was assumed to run under temperature 1000℃ and oxygen utilization ratio was lower than 0.8. At the left of position A, the stack was operated at oxygen utilization ratio 0.8 and temperature lower than 1000℃(fig.7.1). At the right of

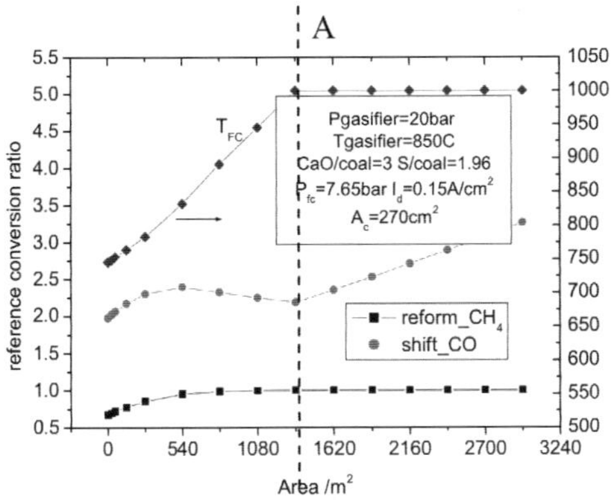


Fig.7.1 Relative conversion ratio and stack temperature vs. area

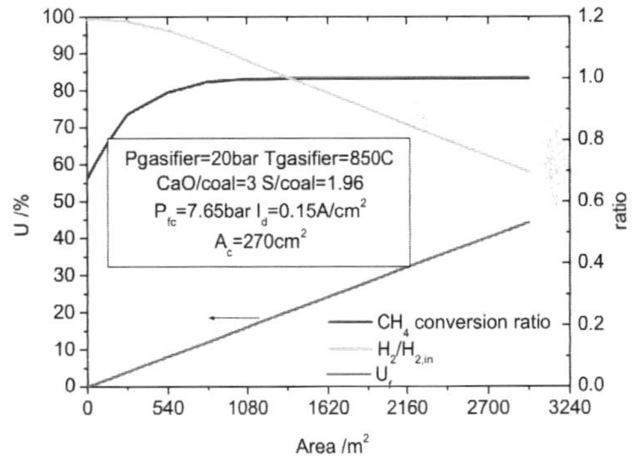


Fig.7.2 Fuel utilization ratio, $H_2$ relative product vs. area

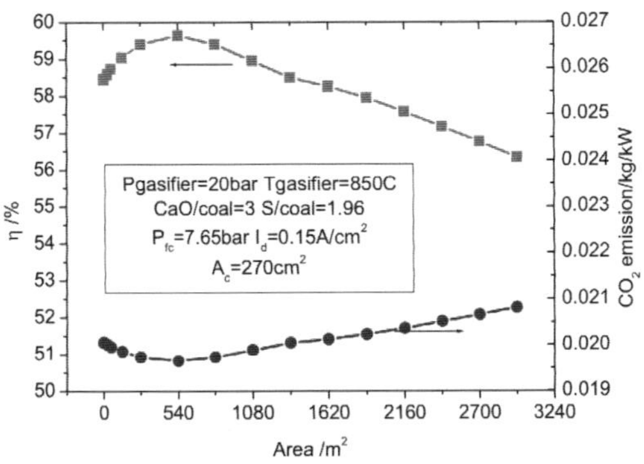


Fig.7.3 Total energy efficiency and equivalent $CO_2$ emission vs. area

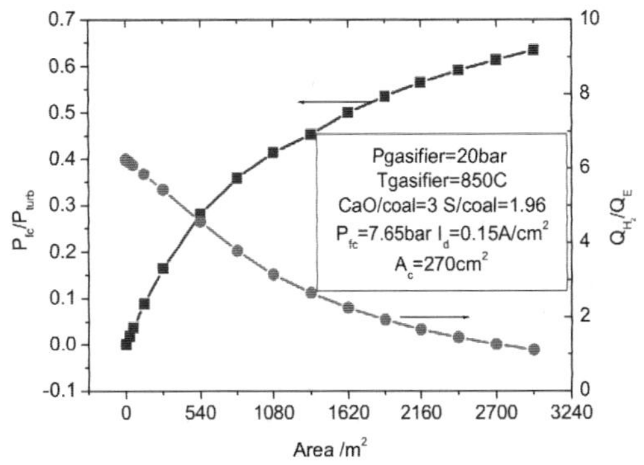


Fig.7.4 Product ratio vs. area

position A, more excess air had to be fed so as to keep temperature not higher than 1000℃ (fig.7.1).. With the increase of reaction area, the $CH_4$ was converted quickly and then after position A, the $CH_4$ was almost converted (fig.7.2). If reaction area was low enough, $H_2$ production was changed slowly due to the prominent compensation of reforming reaction(fig.7.2). Figure 7.1 also showed that CO reacted in shift reaction was higher than inlet CO flow This was because additive amount of CO was produced by methane reforming reaction.

Found in fig 7.3, there was a point at which total energy efficiency was highest. This point corresponded to the position at which $CH_4$ conversion was almost finished. Because almost all $CO_2$ generated during the $H_2$ and electricity co-production process can be capture at pure state, the equivalent green house gas emission[13] was mainly produced during coal mine exploitation, coal cleaning and transportation. Thus the green house emission per unit product was contrary to the variation of efficiency. From the point of view of efficiency and emission, it was better to chose lower reaction area at which about 80% $CH_4$ was converted.

2) effect of gasification temperature and pressure

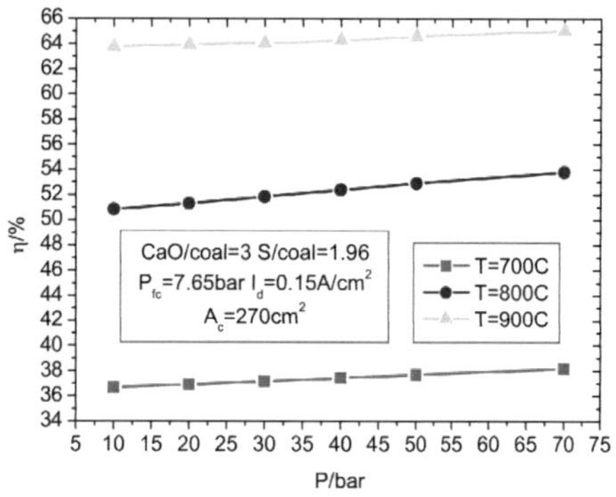


Fig.8.1 Total energy efficiency vs. gasification pressure and temperature

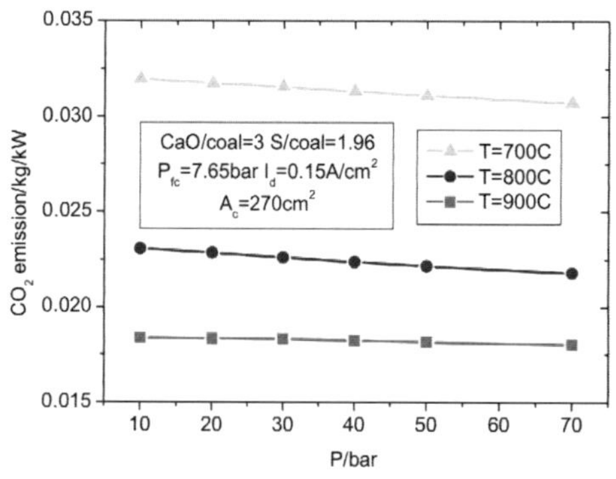


Fig.8.2 Equivalent $CO_2$ emission per unit product vs. gasification pressure

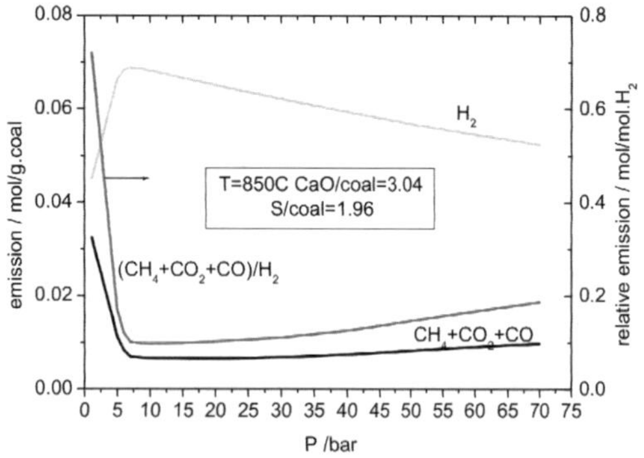


Fig.9.1 Emission per unit $H_2$ vs. pressure

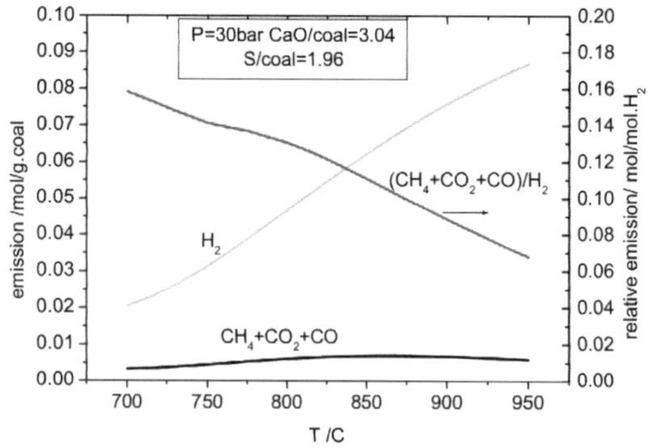


Fig.9.2 Emission per unit $H_2$ vs. temperature

Syngas flow and composition was determined by gasification temperature and pressure. Also by product steam varied with them. So system performance would change with gasifier operation condition. Analysis showed that increase temperature was advantage to increase total energy efficiency and to decrease equivalent GHG(green house gas) emission per unit product of a coal life cycle(fig.8.1 and fig.8.2). The figure9.1 and 9.2 investigated equivalent green house gas emission of gasification process. At temperature 850℃, in the range of 5-30bar, GHG emission changed very little. They showed that higher temperature and lower pressure were more suitable for this section. But to a coal life cycle, higher pressure was more suitable. Apart from efficiency and emission, choice of pressure should also consider the pressure match between syngas and its using facility. Reasonable temperature should also consider the eutectic problem of sorption, heat balance of reactor, etc.

## Conclusions

- Simulation based on chemical kinetics can show the basic characteristics of $CO_2$ sorption enhance gasification. Results illustrated the influence of temperature and pressure on gasifier performance. But the prediction error was still high, especially to low content component. The simulation needs to be improved by adopting more comprehensive hydrodynamic model and correction on kinetics model.
- From gasification chemical kinetics simulation, it was concluded that, if evaluated by $H_2$ yield and composition, and coal to $H_2$ efficiency, increase temperature was advantage both to a coal life cycle(from coal mine exploitation, cleaning , transportation and utilization) and to a separate gasifier section. Lower pressure around 5-30bar was better for gasification section. But to a coal life cycle, higher pressure was more suitable. But the increase of temperature should consider eutectic problem, decomposition of calcium carbonate, heat balance of reactor, etc. Choice of pressure should also consider the pressure match between syngas and its using facility.
- Change reaction area was one way to adjust fuel utilization rate, methane conversion ratio and power output. If reaction area was low enough, $H_2$ production was changed slowly due to the prominent compensation of reforming reaction. With electrochemistry reaction, steam reforming of methane can proceed easily. From the point of view of efficiency and emission, it was better to chose lower reaction area at which about 80%$CH_4$ was converted.

## Acknowledges

The authors would like to thank the supported by the National Natural Science Foundation of China (Grant No. 50876012) and National Key Basic Research Program (NKBRP 2007CB210101) founded by MOST.

## References

1. Lin Shiying,et al(2002). Hydrogen production from coal by separating carbon dioxide during gasification. Fuel, 81, pp2079–2085

2.Madhukar R. Mahishi, et al(2008). A novel approach to enhanced hydrogen yield of biomass gasification using $CO_2$ sorbent, Journal of engineering for gas turbine and power, 130,pp011501-1~011501-8.

3. Tobias Pröll, et al(2008). $H_2$ rich syngas by selective $CO_2$ removal from biomass gasification in a dual fluidized bed system — Process modelling approach. Fuel processing technology,89,pp1207–1217.

4. Lin Shiying,et al(2005).$CO_2$ separation during hydrocarbon gasification. Energy ,30, pp2186-2193.

5. Guan Jian(2007). The Mechanism Study of a New near Zero Emissions Coal Utilization System with Combined Gasification and Combustion. PHD thesis. Zhejiang University.

6. Lin Shiying,et al(2004). Continuous experiment regarding hydrogen production by coal/CaO reaction with steam (I) gas products. Fuel, 83,pp869–874.
7. Qiao chunzhen(2006). Study on the $CO_2$ Sorbents in the Single-Step Hydrogen Production from Carbonaceous Energy. PHD thesis. Institute of Engineering Thermophysics Chinese Academy of Sciences.
8. M. L. de Souta-Santos(1989). Comprehensive modelling and simulation of fluidized bed boilers and gasifiers. FUEL, 68, ,pp1507-1521.
9. S Sundaresan And Neal R Amundson（1979）. Studies In Char Gasification-I a Lumped Model. Chemical engineering science, 34， pp345-354.
10. David Merrick(1983). Mathematical models of the thermal decomposition of coal 1. The evolution of volatile matter. FUEL, 62, May,534-539.
11.A. P. Watkinson, J. P. Lucas and C. J. Lim(1991). A prediction of performance of commercial coal gasifiers. FUEL, ,70, pp519-527.
12. Wang feng（2007）. Study of Direct Hydrogen Production from Carbonaceous Fuels. PHD thesis. Institute of Engineering Thermophysics Chinese Academy of Sciences.
13. Xia Dejian（2010）. Measurement of Life-Cycle Carbon Equivalent Emissions of Coal-Energy Chain. Statistical Research， 8,pp82-89.

# (624f) Process Design of Enriched Hydrogen Gas Production From Empty Fruit Bunch Via Steam Gasification with in-Situ $CO_2$ Capture

Abrar Inayat, Murni M Ahmad, Mohamed I A Mutalib and Suzana Yusup, Chemical Engineering, Universisti Teknologi PETRONAS, Tronoh, Malaysia

Due to the substantial production of palm oil in Malaysia, there is abundant availability of biomass in the form of agricultural wastes. Studies have shown that biomass steam gasification with in-situ carbon dioxide capture offers good prospects for the production of hydrogen rich gas. This work focuses on the mathematical modeling of the flowsheet design for hydrogen production from oil palm empty fruit bunch (EFB) using MATLAB. The process under focus is steam gasification with in-situ carbon dioxide capture by CaO. The flowsheet model incorporates the reaction kinetics models of the steam gasification of EFB ($C_{3.4}H_{4.1}O_{3.3}$) and carbon dioxide adsorption, and the material balances. The developed model is used as a platform to investigate the effects of process parameters on the production of hydrogen rich gas from EFB using a single-pass fluidized bed gasifier; specifically the effects of temperature, steam/ biomass ratio and sorbent/biomass ratio on the product gas composition, purity and yield of hydrogen in the product gas stream. Based on the results, the maximum hydrogen purity predicted is 71 mole% at 1150 K at outlet of the gasifier unit with the yield of 107.3 g/kg of EFB. The purity can be enhanced to 99.9 mole% using a filter, a scrubber and a pressure swing adsorption unit. The effect of steam/biomass ratio between 0.5 and 3.5 on the process performance is also reported, the hydrogen concentration and yield increase with respect to the steam/biomass ratio. It is observed that the increase in hydrogen yield is more significant when increasing the steam/biomass ratio compared to when increasing temperature, within the selected ranges. Meanwhile, by increasing the sorbent/biomass ratio, the purity of hydrogen increases, however showing less impact on the hydrogen yield. The mass conversion efficiency increases with increasing temperature and is found to be at the maximum at 84.7% at 1100 K. Moreover, the mass conversion efficiency increases by increasing steam/biomass ratio. The results are compared with published literature and showed good agreement. The study provides a useful simulation tool for the design and optimization of a future experimental work.

See more of this Session: Advances in Gasification Research II
See more of this Group/Topical: Topical E: High Temperature Environmentally Sustainable Energy Processes (sessions joint with the Environmental Division)

# AUTHOR INDEX

# AUTHOR INDEX

**American Institute of Chemical Engineers**
3 Park Avenue
New York, New York 10016-5991

CURRAN ASSOCIATES INC
proceedings.com